Essays in
Biochemistry

Essays in Biochemistry

Edited for The Biochemical Society by

P. N. Campbell
Department of Biochemistry
University of Leeds
England

F. Dickens
15 Hazelhurst Crescent
Findon Valley
Worthing, Sussex,
England

Volume 6

1970

Published for The Biochemical Society by Academic Press
London and New York

ACADEMIC PRESS INC. (LONDON) LTD
Berkeley Square House
Berkeley Square
London W1X 6BA

U.S. Edition published by

ACADEMIC PRESS INC.
111 Fifth Avenue
New York
New York 10003

Library of Congress Catalog Card Number: 65-15522
SBN 12-158106-3

Printed in Great Britain by
William Clowes & Sons Limited
London, Colchester and Beccles

Dr. G. D. Greville

Guy Drummond Greville died suddenly at his home in Cambridge on 13th December, 1969. He was a kind, sympathetic, intelligent and industrious man. It was this last mentioned quality which perhaps contributed most of all to his tragically early death. For example, during the last two years of his life he wrote four major review articles, all of which are clear, concise, critical and informative. In these articles he makes reference to well over a thousand papers which range over studies on pyruvate and fatty acid oxidation to the chemiosmotic hypothesis. It was in his character that he would have read carefully every one of these references. In addition to this work he was actively pursuing several lines of research with mitochondria, fixation in electron micro-scopy and the properties of a new protein, calliphorin.

Guy Greville was an undergraduate at Emmanuel College, Cambridge, and was awarded a First Class Honours B.A. in Chemistry in 1929. It was while he was working with Frank Dickens, who has replaced him as co-editor in this volume of *Essays in Biochemistry*, at the Middlesex Hospital Medical School that he became a biochemist. His first work was on carbohydrate metabolism, but the subject of his Ph.D. dissertation, which he did under the supervision of Sir Charles Dodds, was the effect of "uncoupling" agents on tissue respiration. In 1937 he moved to Runwell Mental Hospital to take charge of the Biochemical Laboratories there, and worked on several aspects of the biochemistry of mental disorders.

In 1944 he was appointed University Lecturer in Biochemistry at Cambridge. He was a superb teacher and shone even in competition with Ernest Baldwin, Ernest Gale and David Bell. He became Head of the Biochemistry Department of the Agricultural Research Council Institute of Animal Physiology at Babraham, Cambridge. The loss to Cambridge University, however, was not total, since he was often to be seen in the University Biochemistry Department discussing with a wide circle of friends every aspect of biochemistry.

Guy Greville was keenly anticipating moving into the new Biochemical Laboratories at Babraham, the design and construction of which he had supervised with typical thoroughness, and it is very sad that this was not to be.

British biochemistry has suffered a sad loss, and so have all his many friends. One remembers the many entertaining and amusing conversations, the valued advice of a sympathetic friend. It is hard to realize that he is not still with us.

September 1970   BRIAN CHAPPELL

List of Contributors

K. BREW, *Department of Biochemistry, University of Leeds, Leeds LS2 9LS, England* (p. 93)

T. L. COOMBS, *Natural Environment Research Council, Fisheries Biochemical Research Unit, University of Aberdeen, Aberdeen, Scotland* (p. 69)

E. H. FISCHER, *Department of Biochemistry, University of Washington, Seattle, Washington 98105, U.S.A.* (p. 23)

P. T. GRANT, *Natural Environment Research Council, Fisheries Biochemical Research Unit, University of Aberdeen, Aberdeen, Scotland* (p. 69)

M. KLINGENBERG, *Institut für Physiologische Chemie und Physikalische Biochemie, Universität München, 8000 München 15, Germany* (p. 119)

A. POCKER, *Department of Biochemistry, University of Washington, Seattle, Washington 98105, U.S.A.* (p. 23)

E. RACKER, *Section of Biochemistry and Molecular Biology, Cornell University, Ithaca, New York 14850, U.S.A.* (p. 1)

J. C. SAARI, *Department of Biochemistry, University of Washington, Seattle, Washington 98105, U.S.A.* (p. 23)

Conventions

The abbreviations, conventions and symbols used in these Essays are those specified by the Editorial Board of *The Biochemical Journal* in *Policy of the Journal and Instructions to Authors* (Revised 1970). The following abbreviations of compounds, etc., are allowed without definition in *The Biochemical Journal*.

ADP, CDP, GDP, IDP, UDP: 5'-pyrophosphates of adenosine, cytidine, guanosine, inosine and uridine
AMP, etc.: adenosine 5'-phosphate, etc.
ATP, etc.: adenosine 5'-triphosphate, etc.
CM-cellulose: carboxymethylcellulose
CoA and acyl-CoA: coenzyme A and its acyl derivatives
DEAE-cellulose: diethylaminoethylcellulose
DNA: deoxyribonucleic acid
DNP-: 2,4-dinitrophenyl-
EDTA: ethylenediaminetetra-acetate
FAD: flavin-adenine dinucleotide
FMN: flavin mononucleotide
GSH, GSSG: glutathione, reduced and oxidized
NAD: nicotinamide-adenine dinucleotide
NADP: nicotinamide-adenine dinucleotide phosphate
NMN: nicotinamide mononucleotide
P_i, PP_i: orthophosphate, pyrophosphate
RNA: ribonucleic acid
tris: 2-amino-2-hydroxymethylpropane-1,3-diol

The combination NAD^+, NADH is preferred.

The following abbreviations for amino acids and sugars, for use only in representing sequences and in Tables and Figures, are also allowed without definition.

Amino acids

Ala: alanine	Gly: glycine	Orn: ornithine
Arg: arginine	His: histidine	Phe: phenylalanine
Asn*: asparagine	Hyl: hydroxylysine	Pro: proline
Asp: aspartic acid	Hyp: hydroxyproline	Ser: serine
Cys: cysteine	Ile: isoleucine	Thr: threonine
CyS: cystine (half)	Leu: leucine	Trp: tryptophan
Glu: glutamic acid	Lys: lysine	Tyr: tyrosine
Gln†: glutamine	Met: methionine	Val: valine

* Alternative, $Asp(NH_2)$ † Alternative, $Glu(NH_2)$

Sugars

Ara: arabinose	Glc: glucose
dRib: deoxyribose	Man: mannose
Fru: fructose	Rib: ribose
Gal: galactose	Xyl: xylose

Abbreviations for nucleic acid used in these essays are:

mRNA: messenger RNA
rRNA: ribosomal RNA
tRNA: transfer RNA

Any other abbreviations are given on the first page of the text.

References are given in the form used in *The Biochemical Journal,* except that the last as well as the first page of each article is cited and also the title. Titles of journals are abbreviated in accordance with the system employed in *World List of Scientific Periodicals,* 4th ed. (London: Butterworths, 1963, 1964, 1965).

Enzyme Nomenclature

At the first mention of each enzyme in each Essay there is given, whenever possible, the number assigned to it in *Enzyme Nomenclature: Recommendations (1964) of the International Union of Biochemistry on the Nomenclature and Classification of Enzymes, together with their Units and the Symbols of Enzyme Kinetics* (Amsterdam: Elsevier Publishing Co., 1965); this document has also appeared as vol. 13 (2nd ed.) of *Comprehensive Biochemistry,* ed. by M. Florkin & E. H. Stotz (Amsterdam: Elsevier Publishing Co., 1965). Enzyme numbers are given in the form EC 1.2.3.4. The names used by the authors of the Essays are not necessarily those recommended by the International Union of Biochemistry.

Contents

The Two Faces of the Inner Mitochondrial Membrane

E. RACKER

*Section of Biochemistry and Molecular Biology, Cornell University,
Ithaca, New York 14850, U.S.A.*

I. Introduction

Membranes have a primary and common function of separating compartments and are therefore likely to have some common properties. If the only function of membranes were to separate compartments, a simple structure such as a bimolecular leaflet of phospholipid would suffice. Whatever the structure of the primary membrane has been, it has acquired numerous secondary functions and components in the course of evolution. Complex lipids and proteins have been inserted which impart to the individual membrane properties which often over-shadow those shared with other membranes. Such specific secondary components can serve as useful markers for the membranes, e.g., cytochrome oxidase for the inner membrane and monoamine oxidase for the outer membrane of mitochondria. They may in fact occupy so much space that they can be mistaken for the membrane itself. For example, it was widely assumed that respiratory enzymes such as cytochrome oxidase are integral components of the inner mitochondrial membrane. When membranes lacking cytochrome

Abbreviations: F_1, coupling factor (mitochondrial ATPase); F_2 to F_6, other coupling factors with no known enzymic function; F_o, a membranous preparation from mitochondria conferring oligomycin (or rutamycin) sensitivity to F_1; CF_o, a membranous preparation from mitochondria conferring oligomycin sensitivity to F_1 after addition of phospholipids; F_c (Ref. 22) or "oligomycin sensitivity conferring protein" (OSCP, Ref. 39), a soluble protein required by membranous preparations for conferral of oligomycin sensitivity to F_1. Coupling factors of Sanadi and co-workers (Refs. 36, 37); factor A resembles F_1; factor B, resembles F_2. Q, ubiquinone or coenzyme Q.

oxidase and other respiratory components were isolated by chemical fractionation of beef heart mitochondria[1] or by physiological depletion during anaerobic growth of yeast,[2] the notion that cytochrome oxidase is an obligatory part of the mitochondrial membrane had to be abandoned. Moreover, it emerges from such studies that we should think in terms of a primary membrane, which contains components essential for the maintenance of structure, whereas secondary components can be removed without destruction of the integrity of the membrane. It should be emphasized that we speak here only of membrane morphology. Obviously, a membrane deficient in cytochrome oxidase cannot be functional in oxidative phosphorylation.

Membranes (like most of us) have two sides: one for the outside and one for the inside. Since it appears that asymmetry of organization plays a significant role in the function of the membrane and since each side of the membrane has properties which are determined not only by the intrinsic character of its constituents but also by the different content of the compartment which it faces, it has become imperative to examine this split personality of the individual membrane in greater depth. It is a fact of research life that investigators choose to study first what is more accessible and it is therefore not surprising to find in the literature more information on the properties of the outer side of membranes than on those of the inner side.

Since the character of a membrane may be dominated by secondary features and since knowledge about the constituents of the primary membrane is scarce, it is very difficult to make generalizations. Is there, for example, a universal "structural protein", as has been claimed? The answer is: we do not know. We *do* know that the so-called structural protein isolated by Criddle *et al.*[3] is not such a protein. We also know that the mitochondrial inner membrane contains hydrophobic proteins which may be structural proteins (see p. 9). However, we prefer not to use this term since it has been previously used as a label for what later turned out to be a mixture of denatured proteins. It was recently shown conclusively[4] that one of the major components in preparations of this "structural protein" is denatured mitochondrial ATPase. This is not surprising since the various reagents such as dodecylsulfate, acetone and urea that were used to make these preparations[5] are known to inactivate ATPase and render it insoluble.

The inner membrane of mitochondria has the function of providing cells with biologically useful energy. For this purpose a number of complex proteins are integrated into this membrane. Among them are the catalysts of respiration, which are organized in a chain capable of oxidizing substrates such as NADH or succinate. These enzymes are organized in such a manner that the energy of oxidation can be conserved and transformed into the energy of the terminal pyrophosphate bond of ATP. This transformation is accomplished by a coupling device which consists of an ATPase and several other coupling factors organized on one side of the membrane. The ATPase (F_1) can be recognized morphologically as the inner membrane spheres 85 Å in diameter[1b, 6, 7] which were first discovered in electron micrographs when phosphotungstate was applied as a negative stain to mitochondria.[8] In intact mitochondria, these spheres line the side of the membrane which faces the matrix.

Thus there are two steps in the process of oxidative phosphorylation which are clearly separable: (a) the energy conservation step, and (b) the energy transformation step. The mechanism of the energy conservation step is particularly controversial as will be discussed later.

Breakage of mitochondria by sonic oscillation yields submitochondrial particles which have the characteristic inner membrane spheres on the outside.[6, 9] Because of this location of the inner membrane spheres and because of several other considerations of altered functions, submitochondrial particles obtained by sonic oscillation were recognized as being "inside-out".[10,11] We are actually not dealing with a physical inversion of the membrane, but with an alteration of its environment. This is documented by experiments which have shown[12] that an externally added antibody against cytochrome c inhibits the respiration of mitochondria but not that of submitochondrial particles. If the same antibody is present during the sonic disruption of the mitochondria, it is trapped inside the vesicles and now inhibits respiration. Similar experiments with ferritin will be described later.

Since submitochondrial particles obtained by sonic oscillation are capable of catalyzing oxidative phosphorylation, it has become possible to analyze the properties of the two sides of the inner mitochondrial membrane under functional conditions by comparing the response of submitochondrial particles and mitochondria to various external reagents. Although the efficiency of oxidative phosphorylation in submitochondrial particles can be as great as that in intact mitochondria, doubts have been expressed in some quarters whether studies on submitochondrial particles are pertinent to mitochondrial physiology. These doubts may at times indeed be justified; yet I believe that membranes, like isolated enzymes (which from a physiological viewpoint are also artifacts) should be purified. The preparation of submitochondrial particles represents a first step towards the purification of the inner mitochondrial membrane. Although studies with such artifacts may lead us away from mitochondrial physiology, we seem to have no alternative if we want to explore the mechanism of ATP generation at the level of molecular biology. We should also bear in mind that studies with "intact mitochondria", separated from the rest of the cell, may also lead us astray particularly if conducted with the illusion of physiological reality.

II. Mitochondrial Compartmentation and Membrane Topography

A. EXPERIMENTAL APPROACHES

Mitochondria are complex structures with multiple compartments which are separated by highly differentiated and selective membranes. In order to reach the matrix, the innermost compartment in which many of the metabolic functions of mitochondria take place, cations and anions must traverse these membranes. This problem of compartmentation has been approached by studies of the movement of cations and anions, measured either directly or indirectly, by their osmotic effects, or by their influence on the functional activities of mitochondria. An indirect but rather popular approach is to study the effect of one substrate on the oxidation of another.[13] In some instances it is clear that one substrate can facilitate the entrance of another into an inaccessible compartment (e.g., malate aiding citrate transport[14]). In other instances facilitation of oxidation may be due to other reasons, by either providing a cofactor (e.g., oxaloacetate for acetate oxidation) or by removing an inhibitor (e.g., glutamate stimulating succinate oxidation by removing oxaloacetate). Only when the rates of simultaneous oxidation of two substrates are equal to the sum, can one assume that translocations of the substrates are independent of each other. When the rate is less than the sum, competition may take place at the level of translocation, but it may also take place at the level of the respiratory chain. These and related problems have been discussed in recent reviews[15-17] and it is apparent that in spite of the complexities of the systems considerable insight into the accessibility of the compartments and the properties of the inner mitochondrial membrane has emerged. An interesting study on the translocation of azide[18] has led to the conclusion that the interaction of this inhibitor with cytochrome oxidase takes place in the matrix compartment of mitochondria.

Another approach to the topography of mitochondria is based on the use of histochemical methods. However, the inner membrane of mitochondria has a very complex internal structure mainly due to the numerous foldings or "cristae". In microscopic sections it is often difficult to recognize what is outside and what is inside and there is considerable controversy with regard to the exact location of various enzyme systems.

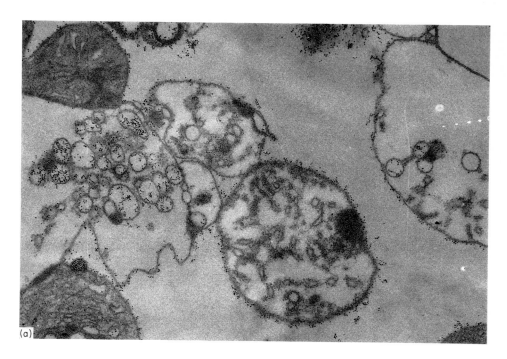

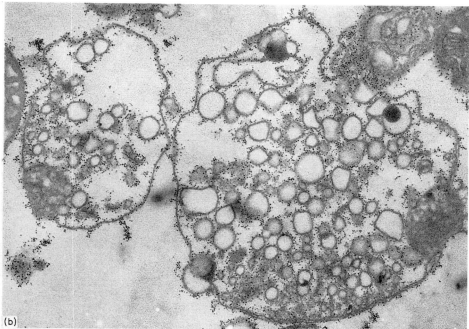

Fig. 1. Electron micrographs of mitochondria exposed to phospholipase C and ferritin. (a) Ferritin present during phospholipase C action. (b) Ferritin added after completion of phospholipase C action.

[Facing p. 4.]

We have recently used a simple procedure to help in this orientation. On exposure of the outer face of the inner membrane to phospholipase C a curious transformation of the mitochondria takes place. The cristae break and form vesicles which have inner membrane spheres on the outside. But in contrast to sonic oscillation, phospholipase C does not rupture the surface of the mitochondria. On exposure of mitochondria to phospholipase C one therefore obtains preparations of submitochondrial particles packaged within mitochondria as shown in Fig. 1. If ferritin is present during phospholipase C treatment, it appears inside the newly formed vesicles (Fig. 1(a)); if ferritin is added after phospholipase C treatment (Fig. 1(b)), it appears outside the vesicles.[12] A schematic drawing illustrating the breakage of the cristae by phospholipase C in the presence of ferritin is shown in Fig. 2.

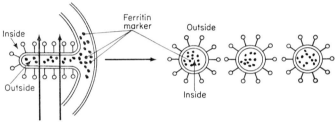

FIG. 2. Scheme illustrating action of phospholipase C on mitochondria. Arrows indicate points of breakage of cristae within the mitochondria.

A histochemical method has been developed presumably for the location of cytochrome oxidase.[19] It was concluded from data obtained by this method that the enzyme is located on the outer side of the inner membrane. The authors, however, clearly recognized that the electron-dense deposits which are formed on oxidation are not a stain for cytochrome oxidase *per se* but for the site of reduction and oxidation of cytochrome *c*.

It will be shown later that the apparent discrepancy between the conclusions on the location of cytochrome oxidase based on inhibition experiments with azide[18] and the conclusions reached in view of histochemical experiments[19] is readily resolved by studies on the location of cytochrome oxidase by macromolecular probes. These probes which we have used to study the two faces of the inner membrane are mainly large enzymes or antibodies which cannot penetrate through the membrane. The most informative experiments were conducted with antibodies against purified components of the inner membrane and with hydrolytic enzymes such as trypsin and phospholipase C.

B. TOPOGRAPHY OF THE INNER MITOCHONDRIAL MEMBRANE

Based on experiments with macromolecular probes a tentative topography of the inner mitochondrial membrane can be outlined as shown in Fig. 3.

The side of the membrane to which mitochondrial ATPase (F_1) is attached is referred to as M-side, because it faces the matrix in mitochondria and the medium in submitochondrial particles. We prefer this designation over the term "inner side" because of the obvious

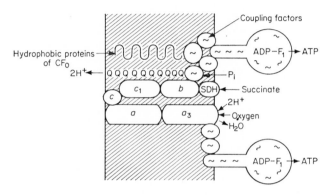

Fig. 3. Topography of the inner mitochondrial membrane. The membrane surface containing cytochrome c is called the C-side, the surface containing coupling factors and succinate dehydrogenase is called the M-side.

confusion arising from the fact that in submitochondrial particles this side faces the outside medium. For the same reason we do not speak of the "outer side" of the inner membrane, but refer to it as the C-side, because it is the physiological location for cytochrome c. I emphasize the physiological aspect of this location because of the curious fact that cytochrome c can also interact with the M-side of the inner membrane as documented by the well-known observation that submitochondrial particles oxidize ferrocytochrome c. The energy of this oxidation cannot, however, be harvested for ATP generation, in contrast to the oxidation of ferrocytochrome c on the C-side of the inner membrane. The simplest explanation for this apparent double-life of cytochrome c is a trivial one. It is based on the finding to be discussed soon that cytochrome oxidase is located on both sides of the inner membrane. As shown in Fig. 3 we place cytochrome a on the C-side and cytochrome a_3 on the M-side of the membrane, but in order to explain the oxidation of ferrocytochrome c on the M-side we make the assumption that cytochrome a_3 as well as cytochrome a can accept electrons from reduced cytochrome c. Such a vectorial organization with cytochrome a on the C-side and cytochrome a_3 on the M-side has appeal in terms of the chemiosmotic hypothesis of Mitchell,[11] but alternative models are easily constructed.

Of the other respiratory catalysts only the location of succinate dehydrogenase has been determined. In view of experiments on the reconstitution of both oxidation, oxidative phosphorylation and respiratory control by addition of succinate dehydrogenase to submitochondrial particles,[20, 21] there can be little doubt regarding its location on the M-side of the membrane. Cytochrome b and c_1 have been tentatively placed in Fig. 3 within the membrane, partly because of the difficulties encountered to separate these components from the membrane and partly to allow them to function as respiratory mediators beween succinate dehydrogenase on the M-side and cytochrome c on the C-side. Since a variety of experiments suggest that the interaction of cytochrome a_3 with oxygen takes place on the M-side, cytochrome a becomes the natural candidate for establishing the continuity between cytochrome c and a_3. We therefore visualize the respiratory carriers of succinoxidase to be organized in a loop from the M-side to the C-side and back to the M-side. It is a single loop rather than the double loop postulated by Mitchell.[11] But it would not be too difficult to modify the topography to make two loops instead of one. Alternatively (as shown in Fig. 3) the protons from succinate may traverse the membrane vectorially to the C-side via the mobile carrier system of ubiquinone (Q).

Linked to the respiratory chain is a coupling device consisting of several coupling factors including F_1 (mitochondrial ATPase). They are associated with the hydrophobic proteins of CF_0 which may represent the structural backbone of the membrane. The hydrophobic proteins in CF_0 which are required for the conferral of oligomycin sensitivity to mitochondrial ATPase are very heat-labile and sensitive to trypsin. Addition of coupling factors to CF_0 stabilizes these proteins remarkably against inactivation by either heat or trypsin.[22] These observations represent another example for allotopy,[23] a phenomenon observed with several membrane components which undergo profound modification of their properties when they are separated from the membrane.

The location of the hydrophobic proteins is shown in Fig. 3 inside the membrane but this is a tentative allocation unsupported by experimental evidence. The location of coupling factors on the M-side of the membrane is well supported by experiments on the resolution and reconstitution of oxidative phosphorylation and in the case of F_1 by functional tests with an antibody against F_1.[12]

Finally, there are phospholipids present in the membrane. As judged by the susceptibility to phospholipases, these phospholipids are present also on both sides of the membrane. As much as 70% of the total phospholipid can be cleaved by exposing both sides of the membrane to phospholipase C without damaging the potential capacity of the particles to catalyze oxidative phosphorylation.[12] This was accomplished by exposure of mitochondria to phospholipase C followed by isolation of submitochondrial particles which were again treated with phospholipase C. Even prolonged exposure to large amounts of phospholipase C did not result in the release of more than 70% of phospholipid phosphorus indicating that about 30% of the phospholipids are either inaccessible from the surface or not susceptible to phospholipase C. This core phospholipid which is undigestible by phospholipase C probably must play an important role in oxidative phosphorylation since phospholipids are required for the reconstitution of succinoxidase[23] and of oligomycin-sensitive ATPase.[1a, 22]

III. The Oxidation Chain

The formulation of the oxidation chain shown below is based on experiments on the reconstitution of succinoxidase from individual components.[24]

Succinate $\rightarrow$ flavoprotein $\rightarrow$ Q-cytochrome b $\rightarrow$ cytochrome c_1 $\rightarrow$
cytochrome c $\rightarrow$ cytochrome oxidase

In the corresponding scheme of NADH oxidation shown below the formulation of the first segment is based on the experiments of Hatefi *et al.*[25] on the reconstitution of NADH-Q reductase from individual components:

NADH $\rightarrow$ flavoprotein $\rightarrow$ nonheme iron protein $\rightarrow$ Q-cytochrome b $\rightarrow$
cyto c_1 $\rightarrow$ cyto c $\rightarrow$ cytochrome oxidase

Some qualifications should be made with respect to these schemes. They are minimal formulations, i.e., the listed components appear to suffice for reconstituting an oxidation chain which resembles that of phosphorylating particles in its sensitivity to rotenone, antimycin and KCN. Moreover, the specific activity of the rate of oxidation catalyzed by the reconstituted complex is comparable to that of phosphorylating submitochondrial particles.

The reconstituted succinoxidase complex was, however, not capable of oxidative phosphorylation even on addition of all known coupling factors. It is conceivable that additional membrane components, e.g., the non-heme iron protein described by Rieske et al.[26] may be required for a functional phosphorylating oxidation chain.

The preparation of cytochrome b used in the reconstitution of succinoxidase[24] was insoluble. It was obtained from a particle preparation enriched in cytochrome b and c_1 by cleavage with guanidine and removal of solubilized cytochrome c_1. Attempts to solubilize cytochrome b without loss of reconstitutive activity failed. Although the preparation was virtually free of all other known members of the respiratory chain as well as of non-heme iron protein, it was obviously not pure. Recent experiments by Dr. C. Cunningham in our laboratory have revealed the presence of several hydrophobic proteins in these preparations, but none of them appeared to be related to core protein[27] when analyzed by polyacrylamide gel electrophoresis. The cytochrome b preparation was also lacking at least some of the functional hydrophobic proteins of CF_0,[22] as indicated by its inability to confer oligomycin sensitivity to added F_1. The preparation of cytochrome c_1 obtained by cleavage with guanidine is reconstitutively active. A preparation of cytochrome c_1 isolated by a harsher procedure[28] was usually inactive in reconstitution experiments. Experiments with submitochondrial particles that have been depleted of succinate dehydrogenase by exposure to alkali, have shown that succinate dehydrogenase isolated in the presence of succinate is reconstitutively active[29] while the dehydrogenase isolated without succinate is reconstitutively inactive in spite of the fact that it catalyzes a rapid reduction of artificial acceptors by succinate. Similar observations were made in experiments on the reconstitution of a short segment of the respiratory chain between succinate and Q.[30] An exploration of the structural differences of succinate dehydrogenase isolated either in the presence or absence of succinate should be a most interesting project.

These studies on the reconstitution of succinoxidase show that special precautions must be taken during the purification of the individual components so that their ability to interact with each other and with phospholipids is not damaged. The approach to the purification of membrane-bound enzymes requires therefore a new attitude with regard to both methodology and evaluation of the product. A protein chemist who purifies a soluble enzyme can rely on an assay of the specific activity of the enzyme and he is not, and should not be, satisfied until he has isolated a homogeneous protein suitable for kinetic and physical chemical studies. An investigator of a membrane-bound enzyme on the other hand, cannot be satisfied by an assay of specific activity alone, since he must also consider parameters of interaction with other membrane components. It may therefore not always be possible to apply to these proteins the same rigid criteria of homogeneity if purity cannot be achieved without loss of reconstitutive activity.

Unfortunately, even the capacity to reconstitute, though essential, does not necessarily suffice for proper functioning of the membrane components. As mentioned above, the reconstituted succinoxidase complex catalyzed oxidation without phosphorylation even after addition of all known coupling factors. As will be postulated later, one of the possible reasons for this failure is the lack of vectorial organization within the membrane. In these earlier experiments,[24] cytochrome c was added externally to the complex and no systematic attempts were made at the time to include considerations of membrane topography in the reconstitution. Subsequently, the importance of vectorial organization became apparent.[12] Methods had to be explored that would allow the respiratory catalysts to assemble in the order found in phosphorylating submitochondrial particles.

Our first attempts in vectorial reconstitution were directed toward an oriented replacement of cytochrome c. A simple tool was available to evaluate the success of these experiments since we have found that antibody against cytochrome c does not inhibit NADH or succinate oxidation in submitochondrial particles which are inside-out.[12] This indicated not only that the location of cytochrome c is indeed on the C-side of the membrane which is inaccessible from the outside in submitochondrial particles, but also a lack of fluidity of cytochrome c and of the antibody within the membrane. These conclusions were supported by two experiments. Complete inhibition of oxidation of NADH or succinate was observed in submitochondrial particles that were prepared by exposure of mitochondria to sonic oscillation in the presence of antibody against cytochrome c. Particles prepared in the presence of normal rabbit γ-globulins were used as controls. Another approach to the same problem was less dramatic, but proved more useful because of its reversibility. Submitochondrial particles that were treated with low concentrations of deoxycholate or with cholate in the presence of salt, were rendered sensitive to the antibody against cytochrome c. The same procedure could be used to remove cytochrome c from these particles as well as to restore cytochrome c to submitochondrial particles that have been depleted of cytochrome c. After removal of the detergents by washing, such restored particles exhibited insensitivity towards the antibody. Addition of deoxycholate again induced sensitivity showing that the cycle of closing and opening of the membrane could be repeated without damage to respiratory chain.[12] However, the procedure caused some damage since oxidative phosphorylation was seriously impaired even after reconstitution with coupling factors. Thus the appropriate vectorial orientation of cytochrome c may be a necessary but not sufficient feature of organization of the membrane required for energy conservation during respiratory events. That the energy-conservation phase rather than the coupling device was affected, was indicated by experiments carried out by Dr. A. Knowles who found marked losses in the fluorescence enhancement of 8-anilino-1-naphthalene sulphonic acid[31] during succinate oxidation even after mild treatment with detergents. More gentle procedures to restore cytochrome c to the C-side of the membrane are now being explored in our laboratory.

The orientation of cytochrome oxidase in the membrane is controversial, as mentioned previously. Electron microscopy places it on the C-side of the membrane;[19] inhibition studies with azide[18] indicate its location at the M-side of the membrane. Experiments in our laboratory with macromolecular probes show that both conclusions are correct. The interaction of cytochrome oxidase (cytochrome a) with cytochrome c indeed takes place at the C-side of the membrane. However, cytochrome oxidase is also present on the M-side of the membrane where cytochrome a_3 interacts with oxygen. Two macromolecular reagents, an antibody against cytochrome oxidase[12] and polylysine, which inhibit the activity of cytochrome oxidase,[32] were shown to diminish effectively the oxidation of ascorbate-cytochrome c in mitochondria as well as in submitochondrial particles. It should be stressed again that the oxidation of ascorbate via external cytochrome c and cytochrome oxidase is coupled to phosphorylation only in mitochondria but not in submitochondrial particles. Since the reduction of cytochrome a is presumed to take place at the C-side of the membrane, it is puzzling to find that submitochondrial particles oxidize externally added reduced cytochrome c. Since we proposed (Fig. 3) that cytochrome a_3 is located on the M-side of the membrane, we must therefore assume either that cytochrome a_3 can also react with cytochrome c or that both a and a_3 are present at the surface of submitochondrial particles. The Mitchell hypothesis requires a vectorial organization of cytochrome a and a_3 as outlined in Fig. 3, and readily explains the lack of phosphorylation associated with the oxidation of reduced cytochrome c in submitochondrial particles

Experiments on the reconstitution of cytochrome oxidase to particles which were completely free of this enzyme have been recently conducted in our laboratory.[32a] The results show that a functional reconstitution of cytochrome oxidase may be achieved together with that of cytochrome c. In such reconstituted particles the fraction of oxidation which was insensitive to the antibody against cytochrome c was coupled to phosphorylation.

IV. The Coupling Device

When surface components are removed from submitochondrial particles by physical or chemical methods the ability to phosphorylate is usually lost before there is any detectable damage to the oxidation chain. This procedure, which we refer to as "resolution from without", gives rise to particles that oxidize NADH or succinate without coupled phosphorylation. On addition of specific membrane components, called coupling factors, phosphorylation can be restored. At least five coupling factors have been described which appear to be distinct. F_1, a homogeneous protein which catalyzes the hydrolysis of ATP, is the only coupling factor with known enzymic activity. This ATPase activity is insensitive to oligomycin but can be rendered sensitive by recombination of the enzyme with the inner mitochondrial membrane. The other coupling factors (F_2, F_3, F_5 and F_6) are also proteins but with no known enzymic function. It remains to be seen whether they play a role as catalysts or as structural components of the membrane. Indeed, alternative functions should be considered. Some of these functions may either be regulatory, e.g., a counteraction of a naturally occurring inhibitor of phosphorylation, or conservatory, e.g., similar to that of oligomycin.[33, 34] Such a conservatory role has been proposed for F_2.[35]

Sanadi and his collaborators[36, 37] have described two coupling factors designated A and B. Factor A is very similar to F_1 and can be replaced by it. The most highly purified preparation of factor B* is clearly different from F_1, F_3, F_5 and F_6. We find that preparations of F_2 substitute for factor B, in stimulating the ATP-driven reduction of NAD by succinate in particles deficient in factor B.[37] Highly purified preparations of factor B substitute for F_2 in stimulating the $^{32}P_i$-ATP exchange assay designed to test F_2. Antibody against factor B eliminates the stimulation by either factor B or by F_2 in both assay systems. We propose therefore that factors B and F_2 are identical. F_3 and F_5 may be related to each other[38] although they can be distinguished in specific assays. None of the other coupling factors described in the literature have survived the test of time as being distinct entities.

The coupling device of the inner membrane also contains some hydrophobic proteins which have been isolated in particulate form (CF_0). If CF_0 is mixed with phospholipids, F_1 and F_c[22] or 'oligomycin-sensitivity

---

* We wish to thank Dr. D. R. Sanadi and Dr. K. W. Lam for a preparation of highly purified factor B and of an antiserum against factor B.

conferring protein' (OSCP),[39] a complex with oligomycin-sensitive ATPase activity is reconstituted. All phosphorylating particles either have a manifest oligomycin-sensitive ATPase or can be activated to exhibit this activity. It is generally assumed that the membrane-bound ATPase is a component of the coupling device which generates ATP from ADP and inorganic orthophosphate. In the chemiosmotic hypothesis it fulfills this function by participating in transmembraneous proton translocation. In the chemical hypothesis it acts as a trans-phosphorylating agent to form ATP from $X \sim P$ and ADP. In the scheme proposed by Mitchell[11] the formation of ATP from $X \sim Y$ $(X \sim I)$ takes place in a concerted reaction. Accordingly, the contribution of coupling factors is more likely to be structural than catalytic. In the chemical hypothesis which includes several intermediates, coupling factors may play a catalytic role in their transformations.

In view of the central role assigned to the mitochondrial ATPase in both hypotheses, some of the relevant properties of F_1 are briefly summarized.

1. Soluble F_1 is an active ATPase which contains no lipids and requires no lipids for activity. It was shown by a variety of experiments[40] that F_1 has a dual role: a catalytic function which is eliminated when the ATPase activity is destroyed by various chemical treatments, and a structural function which is present in chemical derivatives of F_1 which lack catalytic function. The catalytic function can be demonstrated in submitochondrial particles which are completely resolved with respect to F_1 and which require catalytically active F_1 for phosphorylation. The structural function of F_1 can be demonstrated in partially resolved particles which can be stimulated either by catalytically active or inactive F_1.

A small molecular protein was isolated from mitochondria by Pullman and Monroy[41] which inhibits the ATPase activity of F_1. It was shown recently[42] that Mg^{2+} and ATP are required for this inhibition. The inhibitor–F_1 complex still serves as a coupling factor. On the other hand, treatment of F_1 with trypsin or chymotrypsin destroys both the catalytic and the structural function of this coupling factor without damaging the ATPase activity. Moreover, trypsin-treated F_1 still interacts with submitochondrial particles as indicated by binding studies and the conferral of oligomycin sensitivity.[42] It is clear from these experiments that reactivity with water (ATPase activity) is not an essential requirement for coupling activity of F_1. On the other hand, oligomycin sensitivity appears to be a required, though not sufficient, feature of the membrane-bound F_1 for its proper participation in energy coupling. Thus, trypsin-treated F_1 does not catalyze energy coupling, yet it catalyzes ATP hydrolysis which is sensitive to the small molecular mitochondrial inhibitor as well as to antibody against F_1. It also can still be rendered oligomycin sensitive by recombination with particles.[42]

2. When soluble F_1 is added to a preparation of the inner mitochondrial membrane which has been depleted of phospholipids, the ATPase activity is inhibited. This inhibition is not due to the small molecular protein discussed above, but to another protein referred to as F_c[22] which, like oligomycin, inhibits only membrane-bound ATPase. In fact, it is probable that the inhibitory action of oligomycin is mediated by F_c for the following reasons: F_c[22] or OSCP[39] restore oligomycin sensitivity to particles that had lost it upon sonication at pH 10·4.[43] Moreover, F_1 can attach to particles that have been depleted of both F_c and phospholipids without losing its catalytic activity. On addition of F_c, ATPase activity is inhibited but can be restored by addition of phospholipids.[22] It is therefore quite possible that oligomycin inhibits the membrane-bound ATPase by interfering with the activation of the ATPase–F_c complex by the phospholipids present in the membrane.

3. The active center of the ATPase appears to be located in the hydrophobic interior of the protein. Proteolytic enzymes which attack the surface of F_1 do not inhibit its ATPase activity. In contrast, treatment of F_1 with iodine or several other reagents results in inactivation of ATPase activity without losing its ability to attach to the membrane and to serve as a structural component as measured by the stimulation of oxidative phosphorylation in particles partly depleted in F_1 (cf. Ref. 40).

4. Soluble F_1 appears in electron micrographs as a spherical structure of 85 Å in diameter with no visible attachments. Submitochondrial particles which have been depleted of F_1 show a relatively smooth surface. Occasionally, protrusions resembling stalks can be seen either in F_1 or in membrane preparations; however, they are rare and cannot account for the numerous stalks that are seen in electron micrographs as connecting links between F_1 and the membrane. It is possible therefore that the stalks are not preformed but are extruded during the interaction of F_1 with the membrane or indeed during interaction of the negative stain (phosphotungstate) with membrane-bound F_1. The suggestion that OSCP is the protein responsible for the binding of F_1 to the particles and that it represents the stalk,[44] seems premature in view of the reported finding that particles depleted in F_c (OSCP) readily bind F_1.[22] No striking differences in morphology could be detected between F_1 bound to F_c-depleted particles and control particles.

5. The key to the understanding of the oligomycin-sensitive ATPase

is the role played by the hydrophobic proteins of CF_o. As mentioned previously, these proteins may indeed represent the true structural proteins of the inner membrane. Attempts in our laboratory by Dr. C. Cunningham to identify one of them as the core protein[27] were unsuccessful. Some of these hydrophobic proteins may be gene products of the mitochondrial genetic system, since they are missing in the cytoplasmic petite mutant of yeast.[45] The role of these proteins in the conferment of oligomycin sensitivity to F_1 is readily documented by experiments with trypsin which rapidly destroys the ability of particles to confer oligomycin sensitivity. F_1 protects against the damaging effect of trypsin. Oxidative phosphorylation is also readily inactivated by trypsin in particles that are deprived of F_1 but not in particles that have a full complement of F_1.[46] These and other properties of the membrane-bound ATPase reaffirm the close association of the oligomycin-sensitive ATPase with the process of oxidative phosphorylation.

V. Hypotheses

It was emphasized in the introduction that although the primary function of membranes is to separate compartments, acquisitions of secondary components have expanded the biological function of membranes. It is of particular interest that such secondary features can be lost under physiological conditions that do not affect the viability of the cell. Perhaps the most dramatic example is the loss of the respiratory chain from inner membranes of yeast grown anaerobically and the conversion of the "promitochondria" to mitochondria on admission of oxygen.[2]

The question then arises whether the assembly of the respiratory chain in the inner membrane is merely for the purpose of obtaining the high efficiency of a compact organization or whether the process of oxidative phosphorylation is linked in a compulsory manner to the ability of the membrane to separate compartments. Indeed this is the key question which separates the two current major hypotheses of oxidative phosphorylation.

In the chemical hypothesis a horizontal organization of the oxidation chain and of the coupling device should suffice for the operation of the process of oxidative phosphorylation. According to the chemiosmotic hypothesis a compartmentation which allows for the attainment of a proton gradient and a membrane potential is an essential feature. A score board of the experimental evidence favoring one or the other of these two hypotheses has been presented in a recent review.[40] The most important features of this evaluation can be briefly summarized as follows.

(a) The chemiosmotic hypothesis of Mitchell,[11] which has recently been presented in an excellent review comprehensible to biochemists,[47] has explained a number of puzzling experimental observations that put a considerable strain on the chemical hypothesis. All membrane preparations capable of oxidative phosphorylation are found to be vesicular. Membrane preparations consisting of broken fragments are capable of catalyzing oxidation, but no phosphorylation takes place. Solvents like acetone, that affect the permeability properties of particles without changing their morphological vesicular structure, destroy the phosphorylation system without necessarily impairing the oxidative function of the particles. Moreover, for phosphorylation, some submitochondrial particles require membrane components that fulfill a structural rather than a catalytic role. For example, A-particles which have a high content of catalytically active F_1 must be supplemented either with additional F_1 which need not be catalytically active, or with low concentrations of oligomycin.[48,49] Dr. P. Hinkle has obtained evidence that coupling factors are required for proton translocation. Oligomycin at low concentrations has also been reported[50] to affect proton permeability.

According to the chemical hypothesis, uncoupling agents have been proposed to act by facilitating the hydrolysis of a high-energy intermediate. Their chemical diversity requires a degree of complexity in these interactions which is unattractive and improbable. Mitchell proposed[11] that compounds like dinitrophenol uncouple by permitting protons to traverse the membrane and evidence in favor of this notion is accumulating.

One of the most outstanding features of the chemiosmotic hypothesis is the vectorial organization of the components of the oxidation chain within the membrane. Substrates and oxygen react within the membrane from the matrix side, whereas protons are released by hydrogen carriers on the other side (C-side) of the inner membrane. The organization of the respiratory components described earlier is remarkably consistent with the postulated formulations of Mitchell.[11] Although the picture is still incomplete, it is difficult to visualize the need for such a vectorial organization in terms of the chemical hypothesis.

The combined action of valinomycin and nigericin to produce uncoupling in submitochondrial particles[51,52] is more easily understood in terms of the chemiosmotic hypothesis. According to this formulation, nigericin alone is ineffective because it does not affect the membrane potential; valinomycin alone is ineffective because the collapse of the membrane potential is dependent on the availability of potassium inside the vesicle. Nigericin which facilitates an H^+/K^+ exchange[53] provides the required K^+ by exchange with protons that have been translocated into the vesicles during respiration.

(b) The chemical hypothesis on the other hand, is more attractive in explaining numerous partial reactions that are associated with oxidative phosphorylation. Exchange reactions such as the ADP–ATP exchange and the incorporation of $H_2{}^{18}O$ into P_i and ATP are more readily understood in terms of a phosphorylated high-energy intermediate which is not included in Mitchell's formulation of the operation of the reversible membrane-bound ATPase. The failure to isolate such a phosphorylated intermediate should not be used as an argument against the chemical hypothesis any more than a failure to isolate the X∼I, the intermediate postulated by Mitchell, should be used as evidence against the chemiosmotic hypothesis. On the other hand, it should be emphasized that a phosphorylated intermediate could be included in the chemiosmotic hypothesis without damaging its ingenious primary structure.

The chemical hypothesis is also less restrictive with regard to the linkage of ion movements to the respiratory process. Thus the recent observations concerning Ca^{2+} translocation in submitochondrial particles which are inside out[54] and the stoichiometry of K^+/ATP during potassium translocation,[55] are somewhat less embarrassing to the chemical than to the chemiosmotic formulation.

Finally, discrepancies in the kinetics of proton movements and electron movements have repeatedly been claimed by Chance[56] to be incompatible with the chemiosmotic hypothesis.

It is obvious from these considerations that a clear decision in favor of one of the two hypotheses cannot be made. It is not surprising that a number of compromise formulations have been proposed[40, 57] without clearing the fog. The so-called conformational hypothesis[58, 57] is only a variant of the chemical hypothesis. The compulsory translocation of protons during respiration is the key issue which separates the chemiosmotic hypothesis from all other formulations of the chemical hypothesis. It may be very difficult to decide upon this dividing issue without achieving a complete resolution and reconstitution of the system of oxidative phosphorylation.

VI. Concluding Remarks

The catalysts of succinate oxidation located in the inner mitochondrial membrane can be separated from each other and reconstituted to yield an actively respiring complex. The energy of oxidation, however, is lost in heat and cannot be harvested to generate ATP from ADP and inorganic orthophosphate even when coupling factors are added to the complex.

This failure can be explained by the chemical hypothesis in terms of a deficiency in the coupling device or a defective linkage to the respiratory chain. An alternative explanation is provided by the chemiosmotic hypothesis. A defect in the vectorial organization of the respiratory chain should result in a failure to catalyze proton translocation and coupled phosphorylation.

Examination of the topography of a functionally active inner mitochondrial membrane has revealed the presence of a vectorial organization of the respiratory chain that is not present in the artificially reconstituted complex of succinoxidase. We may therefore require a complete knowledge of the membrane topography and a directed vectorial organization of the components of the membrane before we can hope to reconstitute the system of oxidative phosphorylation.

REFERENCES

1. (a) Kagawa, Y. & Racker, E. (1966). Partial resolution of the enzymes catalyzing oxidative phosphorylation. IX. Reconstruction of oligomycin-sensitive adenosine triphosphatase. *J. biol. Chem.* **241**, 2467–2474.
   (b) Kagawa, Y. & Racker, E. (1966). Partial resolution of the enzymes catalyzing oxidative phosphorylation. X. Correlation of morphology and function in submitochondrial particles. *J. biol. Chem.* **241**, 2475–2482.
2. Criddle, R. S., Paltauf, F., Plattner, H. & Schatz, G. (1969). Identification of mitochondrial inner membranes in anaerobically grown baker's yeast. *J. gen. Physiol.* **54**, 57s–65s.
3. Criddle, R. S., Bock, R. M., Green, D. E. & Tisdale, H. (1962). Physical characteristics of proteins of the electron transfer system and interpretation of the structure of the mitochondrion. *Biochemistry, Easton,* **1**, 827–842.
4. Schatz, G. & Saltzgaber, J. (1969). Identification of denatured mitochondrial ATPase in "structural protein" from beef heart mitochondria. *Biochim. biophys. Acta,* **180**, 186–189.
5. Lenaz, G., Haard, N. F., Silman, H. I. & Green, D. E. (1968). Studies on mitochondrial structural protein. III. Physical characterization of the structural proteins of beef heart and beef liver mitochondria. *Arch. Biochem. Biophys.* **128**, 293–303.
6. Racker, E., Tyler, D. D., Estabrook, R. W., Conover, T. E., Parsons, D. F. & Chance, B. (1965). Correlations between electron transport activity, ATPase, and morphology of submitochondrial particles. In *Oxidases and Related Redox Systems,* Vol. 2, pp. 1077–1101. Ed. by King, T. E., Mason, H. S. & Morrison, M. International Symposium on Oxidases and Related Oxidation-Reduction Systems, Amherst, Mass., July 1964. New York: John Wiley.
7. Racker, E. & Horstman, L. L. (1967). Partial resolution of the enzymes catalyzing oxidative phosphorylation. XIII. Structure and function of sub-

mitochondrial particles completely resolved with respect to coupling factor 1. *J. biol. Chem.* **242**, 2547–2551.

8. Fernandez-Moran, H. (1962). Cell-membrane ultrastructure. Low-temperature electron microscopy and x-ray diffraction studies of lipoprotein components in lamellar systems. *Circulation*, **26**, 1039–1065.

9. Löw, H. & Vallin, I. (1963). Succinate-linked diphosphopyridine nucleotide reduction in submitochondrial particles. *Biochim. biophys. Acta*, **69**, 361–374.

10. Lee, C. P. & Ernster, L. (1966). The energy-linked nicotinamide nucleotide transhydrogenase reaction: Its characteristics and its use as a tool for the study of oxidative phosphorylation. *BBA Library*, **7**, 218–234.

11. Mitchell, P. (1966). Chemiosmotic coupling in oxidative and photosynthetic phosphorylation. Glynn Research Ltd., 1–192.

12. Racker, E., Burstein, C., Loyter, A. & Christiansen, R. O. (1970). The sidedness of the inner mitochondrial membrane. In *Electron Transport and Energy Conservation*. Adriatica Editrice, p. 35.

13. Harris, E. J. & Manger, J. R. (1969). Intersubstrate competitions and evidence for compartmentation in mitochondria. *Biochem. J.* **113**, 617–628.

14. Chappell, J. B. (1963). The oxidation of citrate, isocitrate and *cis*-aconitate by isolated mitochondria. *Biochem. J.* **90**, 225–237.

15. Lardy, H. A., Graven, S. N. & Estrada-O, S. (1967). Specific induction and inhibition of cation and anion transport in mitochondria. *Fedn Proc. Fedn Am. Socs exp. Biol.* **26**, 1355–1360.

16. Chappell, J. B. (1968). Systems used for the transport of substrates into mitochondria. *Br. med. Bull.* **24**, 150–157.

17. Pressman, B. C. (1970). Energy-linked transport in mitochondria. In *Mitochondria and Chloroplasts*, pp. 213–250. Ed. by Racker, E. New York: Reinhold Book Corp. (in press).

18. Palmieri, F. & Klingenberg, M. (1967). Inhibition of respiration under the control of azide uptake by mitochondria. *Eur. J. Biochem.* **1**, 439–446.

19. Seligman, A. M., Karnovsky, M. J., Wasserkrug, H. L. & Hanker, J. S. (1968). Nondroplet ultrastructural demonstration of cytochrome oxidase activity with a polymerizing osmiophilic reagent, diaminobenzidine (DAB). *J. Cell Biol.* **38**, 1–14.

20. Fessenden-Raden, J. M. (1970). Reconstitution of highly resolved particles with succinate dehydrogenase and coupling factors. In *IVth Johnson Foundation Colloquium*. Ed. by Chance, B., Lee, C.-P. & Yonetani, T. New York: Academic Press (in press).

21. Lee, C.-P., Johansson, B. & King, T. E. (1969). Reconstitution of respiratory control of succinate oxidation in submitochondrial particles. *Biochem. biophys. Res. Commun.* **35**, 243–248.

22. Bulos, B. & Racker, E. (1968). Partial resolution of the enzymes catalyzing oxidative phosphorylation. XVII. Further resolution of the rutamycin-sensitive adenosine triphosphatase. *J. biol. Chem.* **243**, 3891–3900.

23. Racker, E. (1967). Resolution and reconstitution of the inner mitochondrial membrane. *Fedn Proc. Fedn Am. Socs exp. Biol.* **26**, 1335–1340.

24. Yamashita, S. & Racker, E. (1969). Resolution and reconstitution of mitochondrial electron transport system. II. Reconstitution of succinoxidase from individual components. *J. biol. Chem.* **244**, 1220–1227.

25. Hatefi, Y. & Stempel, K. E. (1967). Resolution of complex I (DPNH-coenzyme Q reductase) of the mitochondrial electron transfer system. *Biochem. biophys. Res. Commun.* **26**, 301–308.

26. Rieske, J. S., Zaugg, W. S. & Hansen, R. (1964). Studies on the electron transfer system. LIX. Distribution of iron and of the component giving an electron paramagnetic resonance signal at $g = 1.90$ in subfractions of complex III. *J. biol. Chem.* **239**, 3023–3030.

27. Silman, H. I., Rieske, J. S., Lipton, S. H. & Baum, H. (1967). A new protein component of complex III of the mitochondrial electron transfer chain. *J. biol. Chem.* **242**, 4867–4875.

28. Goldberger, R., Smith, A. L., Tisdale, H. & Bomstein, R. (1961). Studies of the electron transport system. XXXVII. Isolation and properties of mammalian cytochrome b. *J. biol. Chem.* **236**, 2788–2793.

29. King, T. E. (1963). Reconstitution of respiratory chain enzyme systems. XII. Some observations on the reconstitution of the succinate oxidase system from heart muscle. *J. biol. Chem.* **238**, 4037–4051.

30. Bruni, A. & Racker, E. (1968). Resolution and reconstitution of the mitochondrial electron transport system. I. Reconstitution of the succinate-ubiquinone reductase. *J. biol. Chem.* **243**, 962–971.

31. Azzi, A., Chance, B., Radda, G. K. & Lee, C. P. (1969). A fluorescence probe of energy-dependent structure changes in fragmented membranes. *Proc. natn. Acad. Sci. U.S.A.* **62**, 612–619.

32. Takemori, S., Wada, K., Ando, K., Hosokawa, M., Sekuzu, I. & Okunuki, K. (1962). Studies on cytochrome a. VIII. Reaction of cytochrome a with chemically modified cytochrome c and basic proteins. *J. Biochem., Tokyo*, **52**, 28–37.

32a. Arion, W. J. & Racker, E. (1970). Partial resolution of the enzymes catalysing oxidative phosphorylation. XXIII. Preservation of energy coupling in submitochondrial particles lacking cytochrome oxidase. *J. biol. Chem.* **245**, 5186.

33. Racker, E. & Monroy, G. (1964). Coupling factors at the three sites of oxidative phosphorylation. In Abstracts, VI International Congress of Biochemistry, X (Bioenergetics), p. 760.

34. Lee, C.-P. Azzone, G. F. & Ernster, L. (1964). Evidence for energy-coupling in non-phosphorylating electron transport particles from beef-heart mitochondria. *Nature, Lond.* **201**, 152–155.

35. Fessenden, J. M., Dannenberg, M. A. & Racker, E. (1966). Effect of coupling factor 3 on oxidative phosphorylation. *Biochem. biophys. Res. Commun.* **25**, 54–59.

36. Andreoli, T. E., Lam, K.-W. & Sanadi, D. R. (1965). Studies on oxidative phosphorylation. X. A coupling enzyme which activates reversed electron transfer. *J. biol. Chem.* **240**, 2644–2653.

37. Lam, K. W., Warshaw, J. B. & Sanadi, D. R. (1967). The mechanism of oxidative phosphorylation. XIV. Purification and properties of a second energy-transfer factor. *Arch. Biochem. Biophys.* **119**, 477–484.

38. Fessenden-Raden, J. M. (1969). Partial resolution of the enzymes catalyzing oxidative phosphorylation. XX. Characterization of ASU-particles. *J. biol. Chem.*, **244**, 6662–6667.

39. MacLennan, D. H. & Tzagoloff, A. (1968). Studies on the mitochondrial adenosine triphosphatase system. IV. Purification and characterization of the oligomycin sensitivity conferring protein. *Biochemistry, Easton*, **7**, 1603–1610.

40. Racker, E. (1970). Structure and function of membranes of mitochondria and chloroplasts. In *Mitochondria and Chloroplasts*, pp. 127–171. Ed. by Racker, E. New York: Reinhold Book Corp.

41. Pullman, M. E. & Monroy, G. C. (1963). A naturally occurring inhibitor of mitochondrial adenosine triphosphatase. *J. biol. Chem.* **238**, 3762–3769.

42. Horstman, L. L. & Racker, E. (1970). Partial resolution of the enzymes catalyzing oxidative phosphorylation. XXII. Interaction between mitochondrial ATPase inhibitor and mitochondrial ATPase. *J. biol. Chem.* **245**, 1336–1344.

43. Kagawa, Y. & Racker, E. (1966). Partial resolution of the enzymes catalyzing oxidative phosphorylation. VIII. Properties of a factor conferring oligomycin sensitivity on mitochondrial adenosine triphosphatase. *J. biol. Chem.* **241**, 2461–2466.

44. MacLennan, D. H. & Asai, J. (1968). Studies on the mitochondrial adenosine triphosphatase system. V. Localization of the oligomycin-sensitivity conferring protein. *Biochem. biophys. Res. Commun.* **33**, 441–447.

45. Schatz, G. (1968). Impaired binding of mitochondrial adenosine triphosphatase in the cytoplasmic "petite" mutant of *Saccharomyces cerevisiae*. *J. biol. Chem.* **243**, 2192–2199.

46. Racker, E. (1963). A mitochondrial factor conferring oligomycin sensitivity on soluble mitochondrial ATPase. *Biochem. biophys. Res. Commun.* **10**, 435–439.

47. Greville, G. D. (1969). A scrutiny of Mitchell's chemiosmotic hypothesis of respiratory chain and photosynthetic phosphorylation. In *Current Topics in Bioenergetics*, vol. 3, pp. 1–78. Ed. by Sanadi, D. R. New York: Academic Press.

48. Lee, C. P. & Ernster, L. (1965). Restoration of oxidative phosphorylation in 'non-phosphorylating' submitochondrial particles by oligomycin. *Biochem. biophys. Res. Commun.* **18**, 523–529.

49. Fessenden, J. M. & Racker, E. (1966). Partial resolution of the enzymes catalyzing oxidative phosphorylation. XI. Stimulation of oxidative phosphorylation by coupling factors and oligomycin; inhibition by an antibody against coupling factor 1. *J. biol. Chem.* **241**, 2483–2489.

50. Mitchell, P. (1967). Proton-translocation phosphorylation in mitochondria, chloroplasts and bacteria: natural fuel cells and solar cells. *Fedn Proc. Fedn Am. Socs exp. Biol.* **26**, 1370–1379.

51. Montal, M., Chance, B., Lee, C. P. & Azzi, A. (1969). Effect of ion-transporting antibiotics on the energy-linked reactions of submitochondrial particles. *Biochem. biophys. Res. Commun.* **34**, 104–110.

52. Cockrell, R. S. & Racker, E. (1969). Respiratory control and K$^+$ transport in submitochondrial particles. *Biochem. biophys. Res. Commun.* **35**, 414–419.

53. Pressman, B. C., Harris, E. J., Jagger, W. S. & Johnson, J. H. (1967). Antibiotic-mediated transport of alkali ions across lipid barriers. *Proc. natn. Acad. Sci. U.S.A.* **58**, 1949–1956.

54. Loyter, A., Christiansen, R. O., Steensland, H., Saltzgaber, J. & Racker, E. (1969). Energy-linked ion translocation in submitochondrial particles. I. Ca^{++} accumulation in submitochondrial particles. *J. biol. Chem.* **244**, 4422–4427.

55. Cockrell, R. S., Harris, E. J. & Pressman, B. C. (1966). Energetics of potassium transport in mitochondria induced by valinomycin. *Biochemistry, Easton,* **5**, 2326–2335.

56. Chance, B., Lee, C. P. & Mela, L. (1967). Control and conservation of energy in the cytochrome chain. *Fedn Proc. Fedn Am. Socs exp. Biol.* **26**, 1341–1354.

57. Chance, B. (1969). In *Electron Transport and Energy Conservation. BBA Library* (in press).
58. Boyer, P. D. (1965). Carboxyl activation as a possible common reaction in substrate-level and oxidative phosphorylation and in muscle contraction. In *Oxidases and Related Redox Systems*, Vol. 2, pp. 994–1017. Ed. by King, T. E., Mason, H. S. & Morrison, M. International Symposium on Oxidases and Related Oxidation-Reduction Systems, Amherst, Mass., July, 1964. New York: John Wiley.

The Structure, Function and Control of Glycogen Phosphorylase*

E. H. FISCHER, A. POCKER and J. C. SAARI

Department of Biochemistry, University of Washington

Seattle, Washington, 98105, U.S.A.

I. Introduction

This essay is concerned with the physical, chemical and enzymic properties of glycogen phosphorylase (EC 2.4.1.1.), an enzyme which plays an essential role in the regulation of carbohydrate metabolism. Over the years and for many obvious reasons, this enzyme has attracted the interest of numerous investigators, with the following main conclusions.

* The authors wish to thank the National Institutes of Arthritis and Metabolic Disease, NIH, United States Public Health Service (AM 07902), the National Science Foundation (GN 5932X) and the Muscular Dystrophy Association of America for support during preparation of this review.

Abbreviations. PLP, pyridoxal 5′-phosphate; *p*-CMB, *p*-chloromercuribenzoate; DTNB, 5,5′-dithiobis-(2-nitrobenzoic acid); UDPG, uridinediphosphoglucose; 5′-P-Pxy, 5′-phosphopyridoxyl; FDNB, fluorodinitrobenzene; EDTA, ethylenediaminetetra-acetic acid; EGTA, ethyleneglycol bis-(β-aminoethyl ether)-N,N'-tetra-acetic acid; G-1-P, glucose 1-phosphate; G-6-P, glucose 6-phosphate; Ser-P, phosphoserine; KAF, phosphorylase kinase activating factor.

(a) From the elegant and fundamental work of Cori and his group,[1] phosphorylase was the first enzyme shown to exist in an active and inactive form (phosphorylase a and b, respectively) with the obvious conclusion that conversion of one form to the other would serve as a means to regulate glycogen metabolism.

(b) It was the first example for which interconversion of the two forms was shown to result from a covalent modification of the enzyme, namely, phosphorylation of an amino acid residue on the protein (Krebs and Fischer[2]).

(c) It was the first oligomeric enzyme that could be reversibly dissociated into monomeric species by substitution of –SH groups (Madsen and Cori[3]).

(d) It was perhaps the first enzyme found to require an effector for activity (AMP), structurally unrelated to the substrates.[4,5] In this instance, however, allosteric activation is superimposed upon the covalent control of the enzyme mentioned above.

(e) Since phosphorylase catalyzes the initial step of a fundamental metabolic pathway (i.e., the utilization of glycogen), regulation of its activity occurs through a highly complex and sophisticated set of reactions. Indeed, several divalent metal ions and nucleotides, at least four or five enzymes, and different hormones (depending on the tissue in which this regulation is being exercised) were found to be involved.

(f) All glycogen phosphorylases contain pyridoxal 5'-phosphate (PLP) which is indispensable for enzymic activity;[6] yet the exact role of this cofactor—whether catalytic or structural—has not been established.[7]

(g) Finally, phosphorylase can be considered a material of choice for those who wish to undertake a comparative study of protein structure and control. It is one of the glycolytic enzymes; since the glycolytic pathway is essentially ubiquitous to all forms of life, its component enzymes will be found in every species, from unicellular organisms to all the complex tissues of higher plants and animals. Furthermore, glycolysis must have evolved at an earlier time than the aerobic pathway of carbohydrate metabolism (citric acid cycle), since it is generally assumed that life on Earth emerged in a reducing atmosphere.[8]

This review will attempt to summarize our present state of knowledge of the structure of glycogen phosphorylase, its enzymic properties and some of the complex mechanisms by which its activity may be regulated. Since the rabbit skeletal muscle enzyme has been the most thoroughly investigated, it will serve as the prototype for this discussion, but the properties of phosphorylases obtained from other tissues or species will also be described. Several comprehensive reviews have appeared on the same subject.[9-12]

II. Molecular Properties of Phosphorylase

A. GENERAL STRUCTURE

Rabbit skeletal muscle phosphorylase was shown to exist in two forms: phosphorylase b, a dimer of mol.wt. 185,000, essentially inactive in the absence of AMP; and phosphorylase a, a tetramer of mol.wt. 370,000, active in the absence of this nucleotide (Fig. 1). Conversion of phosphorylase b to a occurs through phosphorylation of the protein by Mg-ATP in a reaction catalyzed by phosphorylase kinase (Krebs and Fischer[2]); phosphorylase a is converted back to b by phosphorylase

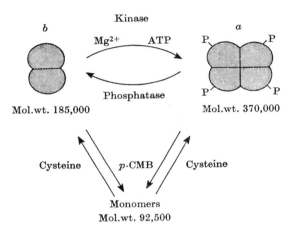

FIG. 1. Molecular forms of muscle phosphorylase.

phosphatase as shown by the Coris and coworkers.[13-15] The properties of these two enzymes which affect the state of activity of phosphorylase will be described in some detail in Section IV dealing with the control of glycogen breakdown. Both phosphorylase b and a can be reversibly dissociated into monomeric subunits (nonphosphorylated units in the case of b and phosphorylated units in the case of a), on treatment with sulfhydryl reagents such as p-CMB,[3] DTNB,[17] or "deforming agents"[18] such as imidazole–citrate (see Section IIC). Dissociation is reversed by removal of the dissociating agent (addition of cysteine in excess in the case of p-CMB or dialysis). Dissociation also occurs in the presence of a number of denaturing agents (urea, guanidine HCl, sodium dodecyl-sulfate, etc.), but in these instances, of course, the process is irreversible. Some of the physical constants of rabbit muscle phosphorylase are summarized in Table 1. Molecular weights of 370,000, 185,000 and 92,500 daltons have been ascribed to phosphorylase a, b, and phosphorylase monomer, respectively.[19] Similar values were obtained by Sephadex chromatography,[20] by the Archibald method of approach to equilibrium in guanidine HCl,[21] and acrylamide gel electrophoresis in sodium dodecylsulfate.[22]

From sedimentation equilibrium studies carried out in 7·2 M guanidine HCl, sodium bicarbonate, acetic acid or concentrated formic acid and other denaturing solvents, no evidence could be obtained that the phosphorylase monomer is made up of more than a single polypeptide chain.[23] A similar conclusion was drawn from acrylamide gel electrophoresis in sodium dodecylsulfate: only one band was obtained from

TABLE 1

Physical parameters of rabbit muscle phosphorylase

Property	Phosphorylase b	Phosphorylase a	References
$S^0_{20, w}$ (s)	$8 \cdot 42 \pm 0 \cdot 03$	$13 \cdot 5 \pm 0 \cdot 1$	19, 20
$D^0_{20, w}$ (10^7 cm^{2}s^{-1})	$4 \cdot 14 \pm 0 \cdot 06$	$3 \cdot 3$	19, 20
$\bar{v}_{20}$	$0 \cdot 737$	$0 \cdot 737$	19
Molecular weight (10^3)	185	370	19–21
Stokes radius (Å)	$49 \cdot 3 \pm 0 \cdot 9$	$63 \cdot 0 \pm 1 \cdot 7$	20
Molecular dimensions (Å) (l × w × h)	$110 \times 65 \times 55$		29, 30

either phosphorylase b or a, corresponding to a material of mol.wt. approximately 95,000 with no evidence for lower molecular weight species. Finally, no evidence was obtained that the monomeric species would be made up of two identical peptide chains linked covalently to one another; since no amino- or carboxy-terminal group was ever detected in rabbit muscle phosphorylase,[24] an end-to-end or even cyclic structure could conceivably be postulated, unlikely as this may be. Against these alternatives are the findings that (a) there is only one phosphorylated site and one PLP-binding site per enzyme monomer;[25, 26] (b) the nine SH groups of the monomeric unit occur in nine peptides of different sequences;[27] and (c) cyanogen bromide attack of the 20 methionyl residues results in the formation of about 21 different peptides.[28]

It has recently been suggested that the two subunits of phosphorylase b may not be identical.[29] This hypothesis was based on the argument that isologous association of identical monomeric units can only give rise to square tetrameric molecules; since rhombic structures were observed by electron microscopy it was proposed that the enzyme is made up of two different subunits. There is no chemical or physical evidence for or against this contention.

Perhaps the most notable characteristic of phosphorylase is its multiplicity of sites, all involved in some way in determining or controlling the activity of the enzyme (Fig. 2). Of course, this is not a feature unique to phosphorylase but rather, one found commonly in regulatory enzymes. First, there is the *catalytic site*, binding the various substrates (glycogen, P_i and G-1-P) and some inhibitors (e.g., glucose or UDPG) and involved in the enzymic reaction. Nothing is yet known on the nature of the residues forming the catalytic site. Second, there is the

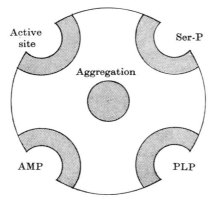

FIG. 2. Schematic representation of sites on phosphorylase *a* monomer.

nucleotide binding site or site of allosteric control of phosphorylase activity. It is the binding of AMP at this site that produces an allosteric change in the conformation[31] of phosphorylase *b* resulting in the appearance of enzymic activity. The same site binds ATP[32, 33] which, therefore, acts as an allosteric inhibitor. Third, there is the *phosphorylated site* involved in the *b* to *a* conversion; phosphorylation freezes the molecule in the active conformation and, therefore, provides for a covalent control of enzymic activity in addition to the allosteric control mentioned above. Fourth, there is the *PLP-binding site* since this cofactor is indispensable to the activity of the enzyme. Finally, there must be several secondary sites or distribution of groups involved in the *subunit assembly* of the molecule; two at least, since one has to account for the formation of the dimer and tetramer forms of the enzyme. Additional groups may be necessary to account for the binding of other effectors such as G-6-P.

Nothing is known about the general topography of the molecule itself, and whether or not some sites might overlap. However, a number of very definite homotrophic and heterotrophic interactions have been recognized to the extent that interaction at any one site appears to affect in some ways the properties of all the other sites and therefore, the general behaviour of the molecule. These will be described in detail in Section III dealing with the kinetic properties of the enzyme.

B. CHEMICAL PROPERTIES

It is obvious that a protein as complex as phosphorylase will not be entirely understood until its detailed structural features are revealed by sequence analysis and X-ray crystallography. That these studies are barely under way reflects a realistic appraisal of the magnitude of the problem involved when dealing with such an enormous molecule. There is now little doubt that the smallest polypeptide chain in phosphorylase contains more than 800 amino acid residues and, therefore, is roughly three times the size of the largest protein so far sequenced (glyceraldehyde-3-phosphate dehydrogenase containing 330 residues).[34]

As if the magnitude of the subunit were not enough, the molecule has a few other subtleties which complicate its chemistry. For instance, end groups for the rabbit muscle protein have never been detected despite the efforts of many investigators.[24] The amino terminus is undoubtedly blocked; however, no acetylated (or, for that matter, formylated) residue could be detected either in the native protein or in the acidic peptides generated by pronase digestion.[28,35] Based on the occurrence of other known amino-terminal blocking groups in mammalian proteins, it appears likely that this residue in phosphorylase is a pyrrolidone carboxylic acid derivative. The latter could be generated by the cyclization of an NH_2-terminal glutaminyl residue or by the introduction of this group at polypeptide chain initiation.[36]

While blocked amino-terminal groups are frequently observed in proteins, the apparent lack of a carboxyl-terminal residue is more unusual. Again, several procedures based on entirely different principles (carboxypeptidase A and B digestions,[37] hydrazinolysis,[38] selective tritiation[39]) gave negative results. The problem is not merely one of size since rat muscle phosphorylase of equal molecular weight was found to possess stoichiometric amounts of isoleucine at its carboxyl end using the same methodology.[40] The only well-documented blocked carboxyl-terminal residue found in a natural peptide is glycinamide present in bovine oxytocin and vasopressin.[41]

Chemical modification studies of muscle phosphorylase have yielded little useful information because of the large number of reactive groups and the inherent lack of specificity of even the so-called "specific" chemical reagents. Only when groups within the protein itself, or particular enzymic reactions, have provided the necessary specificity, have modification studies led to significant results. Thus sodium boro-hydride reduction of phosphorylase proved that PLP could be bound to the enzyme as a Schiff-base[42] and, therefore, that the aldehyde function of this cofactor could not participate in catalysis since the reduced protein was enzymically active.[7] Similarly enzymic phos-phorylation of the protein by phosphorylase kinase was completely specific in that only one serine out of 43 per monomer was sub-stituted.[25, 26]

In an elegant study that opened the way to our understanding of the architecture of oligomeric systems, Madsen showed that muscle phos-phorylase *a* could be dissociated into subunits one-fourth the original molecular weight by reaction of the sulfhydryl groups with the mercurial *p*-CMB.[3, 43, 44] It was noted, for instance, that mercaptide formation preceded inactivation and that dissociation of the protein followed, suggesting the importance of conformational changes in aggregation and activity. Since then, numerous investigators have re-examined the role of sulfhydryl groups on the activity and structure of phosphorylase and arrived at the following conclusions.

(a) Substitution of 1–2 fast reacting SH groups by reagents such as DTNB,[17, 45] FDNB,[46] iodoacetamide,[47, 48] etc., has little effect on the catalytic or oligomeric properties of the enzyme, except that homo-trophic interactions are abolished.[17] In other words, one observes a conversion from allosteric to Michaelian kinetics. These two sulfhydryl groups are associated with the sequences Asn–Glu–Lys–Ile–Cys–Gly–Gly and Gly–Cys–Arg–Asp.[47, 48] (b) A slower reaction with 1–2 additional SH groups results in dissociation of the protein to the monomeric state with complete abolition of enzymic activity; sequences associated with these groups are Asn–Ala–Cys–Asp and Ala–Cys–Ala–Phe. (c) Complete substitution of the nine sulfhydryl groups (the enzyme contains no S—S bond) was only obtained after dissociation of the protein with urea or sodium dodecylsulfate, or after digestion with pepsin.[49] (d) Ligand or covalent induced conformational changes dramatically alter the pattern of sulfhydryl reactivity: for instance, phosphorylase *a* lacks the two slow reacting sulfhydryl groups seen in phosphorylase *b* and thus its activity and state of aggregation are unaffected by reagents such as DTNB.[28] Likewise, the allosteric effector AMP converts the reactivity of the SH groups of phosphorylase *b* into that of phosphorylase *a*.[45]

C. PYRIDOXAL 5'-PHOSPHATE SITE AND ROLE OF THE COFACTOR

Pyridoxal 5'-phosphate, first found as a prosthetic group in a variety of enzymes involved in amino acid transformations,[50, 51] is present in stoichiometric amounts (one equivalent per enzyme subunit) in all phosphorylases so far isolated. Interest in the role of this cofactor was stimulated by the finding that whereas its removal from phosphorylase led to total inactivation of the enzyme,[52] reduction of the protein with sodium borohydride which fixes the cofactor to the protein results in a material which is still 60% enzymically active.[7, 42] All other classical B_6 enzymes in which pyridoxal phosphate is directly involved in catalysis are totally inactivated by this kind of treatment.[51, 53]

One is, therefore, left with the following alternatives. 1. Pyridoxal 5'-phosphate is directly involved in catalysis, as a group carrier, just as in any other typical B_6 enzyme; if so, a functional group other than the 4-aldehyde must be involved. 2. Phosphorylase is a double-headed enzyme capable of catalyzing a second enzymatic reaction (e.g., one of the classical reactions of PLP-containing enzymes) in addition to phosphorolysis. None of these reactions was detected with the pure enzyme.[54] 3. The cofactor is not directly involved in catalysis, but can be used to control the activity of the enzyme. Against this hypothesis are two arguments. (a) Control mechanisms of enzyme activity often vary from one species to another—the activity of a bacterial enzyme is often regulated by effectors which differ from those of a mammalian or a plant enzyme. Yet, PLP has been found in phosphorylase from all species so far investigated: mammalian, avian, crustacean, bacterial, and higher plants. (b) For the activity of phosphorylase to be controlled by means of the cofactor, the latter must be readily accessible on the enzyme; it should be possible either to remove it, modify it, or in some way alter the site at which it is bound. But PLP is very strongly bound to rabbit muscle phosphorylase and behaves as if it were buried within the protein molecule: it will not react with aldehyde reagents unless the protein is modified by changes in pH, or distorted by salts or denaturing agents.[18, 55] 4. Pyridoxal 5'-phosphate is involved neither in direct catalysis nor control but is present as a structural determinant essential for enzymic activity. Indeed, with all its functional groups, PLP can be visualized as a versatile tool to perform this function. However, in terms of cellular metabolism, this is an expensive way to maintain a protein in the proper conformation, and one would wonder why certain species or organisms, after countless mutations, have not been able to find a more conventional way of doing so.

The most intriguing aspect of this problem is that in the rabbit, for instance, there appears to be more vitamin B_6 stored in glycogen phosphorylase than in all the other classical B_6 enzymes taken together.[56]

The multiple roles of this cofactor are no doubt related to its numerous pH-dependent ionic species,[59, 60] each possessing distinct spectral characteristics (Fig. 3). As in other PLP-containing enzymes, the prosthetic group in phosphorylase is covalently linked to an ε-amino residue of a unique lysine, but here the analogy ends. In classical PLP-containing enzymes, the cofactor is linked as a Schiff-base and the

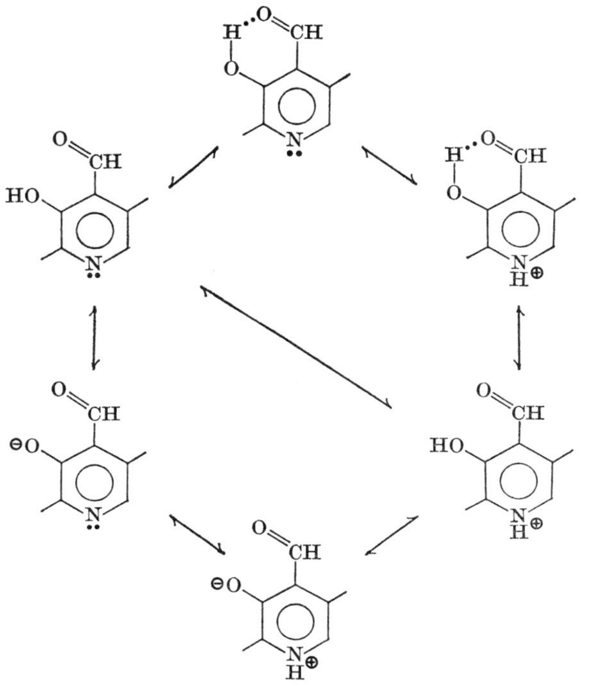

Basic forms	Neutral forms	Acidic forms
max. ~ 390 nm	max. $\begin{cases} 330 \text{ nm} \\ 390 \text{ nm} \end{cases}$	max. ~ 295 nm

FIG. 3. Equilibrium scheme of PLP species in solution including only those forms which primarily contribute to the ultraviolet spectra. Hydration of the carbonyl group and other substituents have been omitted.

participation of this structure in the catalytic mechanism has been amply documented.[50, 61] In every instance, NaBH$_4$ reduction has led to a total loss of activity. Not so in phosphorylase: this enzyme shows a maximum absorption at 330 nm [(I), Fig. 4] in the range of maximal activity (pH 6·5–7·5) that is uncharacteristic of model Schiff-bases. Only when the pH is brought below 4·5 or above 9·5 do phosphorylase solutions become yellow and their absorption shifts to a 415 nm maximum characteristic of a PLP–Schiff-base [(II), Fig. 4]; under these conditions, the enzyme is inactive and unstable. These spectral shifts must result not only from changes in pH but to a great extent from the interaction of the functional groups of the cofactor with their immediate protein environment.

Since no reduction of the enzyme by $NaBH_4$ was observed at neutral pH, a substituted aldamine structure (I) was postulated. Group X could be any nucleophilic residue, not necessarily part of the protein, but it has never been identified. Only after conversion to the Schiff-base (II) does $NaBH_4$ reduction occur to yield the secondary amine (IV) which

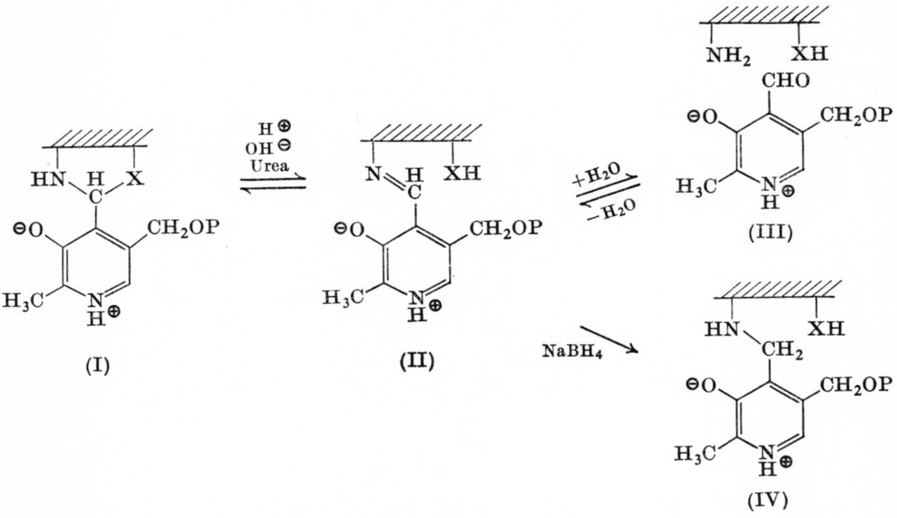

Fig. 4. Sodium borohydride reduction of phosphorylase.

also absorbs at 330 nm. As indicated earlier, this derivative is 60% active and most of its enzymic and physical properties are indistinguishable from those of native phosphorylase b.[7]

In order to characterize further the structure of the pyridoxal 5'-phosphate site, the sequence of amino acids surrounding the substituted lysyl residue was determined. However, this proved to be a difficult task because initially, chymotryptic digestions[25,26] or cyanogen bromide cleavage[28] yielded only small P-pyridoxal peptides. Eventually conditions of limited tryptic and chymotryptic hydrolysis led to the isolation of large fragments; sequential analysis of these taken with the data obtained from the CNBr digestions yielded the structure shown in Fig. 5.[57] This region is remarkably free of basic amino acids with only one occurring in a sequence of some 40 amino acids; statistically one would expect to find a basic amino acid every 7–8 residues.[40] The

Figure also indicates the points of cleavage by various proteolytic enzymes. Trypsin appears to cleave the bond between phenylalanine and methionine with great ease; it is not known if this represents an inherent "chymotryptic-like" activity in trypsin[58] or if it results from the proximity of the P-pyridoxal group.

The prosthetic group in phosphorylase is bound tightly to the protein or "buried" and thus inaccessible to ordinary carbonyl reagents.

FIG. 5. Amino acid sequence of the pyridoxal phosphate-binding site in phosphorylase. For details see text. C, chymotrypsin; T, trypsin; P, pepsin.

Removal of PLP from phosphorylase requires a prior distortion of the molecule that exposes the cofactor;[18, 55] when drastic conditions are used (acidification or denaturing agents) an irreversible loss of activity usually results. However, resolution can be carried out under very mild conditions (pH 6·0, 0°C) if the enzyme is first deformed by specific agents in such a way that the cofactor can now interact with certain carbonyl reagents or even exchange with radioactive PLP added to the medium. A schematic representation of this two-step process is shown in Fig. 6. Deformation is accompanied by dissociation to monomers,

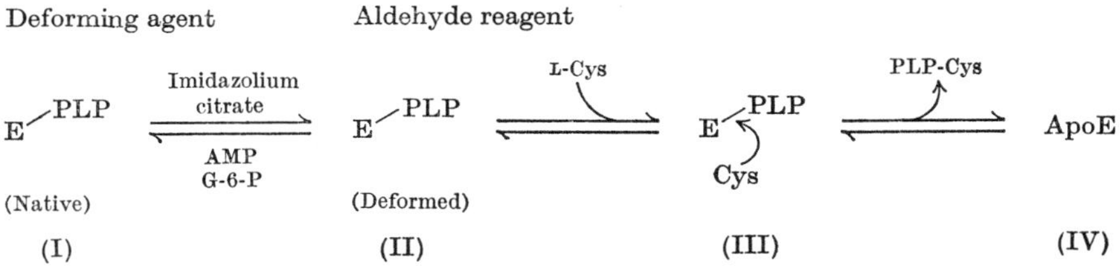

FIG. 6. Scheme for the resolution of phosphorylase.

loss of enzymic activity and of the fluorescence emission at 530 nm due to the bound cofactor. As deforming agents, imidazole and several of its methylated derivatives are highly effective (in contrast to 4-imidazole acetic acid, histidine or histamine), but only when used in conjunction with di-, tri- and tetracarboxylic acids. Imidazole buffers with monovalent acids, e.g., imidazole HCl, are very inefficient.[55]

Resolution is prevented by phosphorylation of the protein (conversion of phosphorylase b to a) or addition of AMP, emphasizing once more the vast changes in conformation that occur under these conditions.[18, 63] Strangely enough, resolution was not only specific towards the deforming agent but, to a very large degree, also towards the nature of the carbonyl reagent. It proceeds with L-cysteine, but not with homocysteine, cysteamine, dimercaptopropanol, etc; most surprising of all, it does not proceed with D-cysteine, indicating that the reaction is stereo-specific.[64]

Addition of stoichiometric amounts of PLP to the apoenzyme restores full activity and the reconstituted protein is indistinguishable from native phosphorylase. The apoenzyme has a high affinity for the cofactor and will preferentially remove this compound when it is mixed with many of its analogs. Reconstitution is enhanced by the presence of L-cysteine and one equivalent of this amino acid is incorporated per enzyme protomer, suggesting that a specific binding site exists. Thus in both resolution and reconstitution, PLP appears to leave or re-enter the enzyme as the thiazolidine derivative of L-cysteine. Reconstitution is strongly temperature-dependent with an energy of activation of 22 kcal/mol.[63–65] Both pyridoxal and 5-deoxypyridoxal (and, for that matter, other aldehydes unrelated to PLP) form a Schiff-base with the enzyme and restore both the state of aggregation and allosteric properties of the native protein, but not its catalytic activity.[63, 67]

In order to assess the possible participation of the functional groups of PLP in catalysis, we undertook a detailed study in which analogs modified in every single position of the pyridine ring[67] were tested for their ability to reactivate apophosphorylase.[63] Figure 7 summarizes the results of this study, indicating that positions 2, 3 and 6 of PLP are not essential for catalysis. A potential aldehyde group in position 4 is necessary for the binding of the cofactor, but not for enzymic activity since sodium borohydride reduction of the enzyme yields an active protein. Most analogs of PLP modified in position 5, i.e., pyridoxal 5'-sulfate or 5'-acetic acid, were found to be inactive, except for the 5'-methylenephosphonate analog[66] which reactivated the enzyme up to about 20%.[68] This may indicate that a group with a pK around neutrality is required for activity (except that yeast glycogen phosphorylase has an optimum pH around 5·8). There is, of course, no exchange of the 5′-phosphate group of PLP with either P_i or G-1-P, two of the substrates of phosphorylase.[52]

Position 1 of PLP has been currently implicated in catalysis: differential spectroscopy of phosphorylase b in the presence of either G-1-P or P_i has revealed a new maximum absorption peak at 360 nm

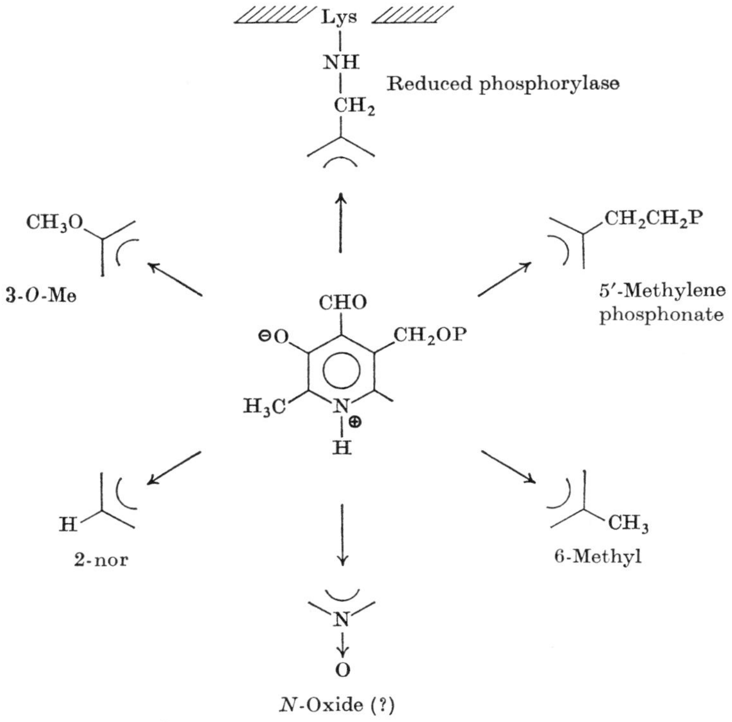

FIG. 7. Active analogs of PLP in phosphorylase.

with a minimum at 330 nm.[69] This was attributed to the formation of an ion pair between the pyridinium nitrogen of PLP and one of the negatively charged groups of the substrates, P_i or G-1-P. Though attractive, this hypothesis does not bear close scrutiny: no 360 nm band is observed with $NaBH_4$-reduced phosphorylase b which is 60% active, nor with 3-O-Me PLP-phosphorylase b which is 25% active. Furthermore, in spite of the fact that substitution at the phenolic group affects the ionization state of the pyridinium nitrogen, no change is observed in the optimum pH of the enzyme reconstituted with

3-O-Me PLP. On the other hand, phosphorylase b reconstituted with 6-methyl PLP (only 8% activity) gives rise to a very strong differential spectrum. This differential absorption, therefore, cannot be correlated simply with the formation of an enzyme–substrate complex; it requires the presence of a potential aldimine and a free phenolic group, and probably reflects some interaction between the cofactor and certain residues on the protein. In this connexion, a charge transfer complex might be implicated. Unfortunately, substitution at the pyridine nitrogen with a methyl group (N-Me PLP) results in an analog which does not bind to the apoenzyme, and, therefore, cannot serve to answer this point.[63]

PLP-N-oxide, on the other hand, appeared to reactivate apophosphorylases b and a up to approx. 50%. However, this analog is unstable in solution and readily regenerates PLP; since this conversion is greatly enhanced by the presence of apophosphorylase, it is not certain whether the observed reactivation is really due to the analog or to the formation of PLP.[68]

D. PHOSPHORYLATED SITE

Phosphorylation of a single seryl residue during the b to a conversion is responsible for the covalent control of phosphorylase activity[2] (see Section IV D). All phosphorylases that are activated in this manner appear to have very similar sequences at this site (Table 2). These sequences bear no resemblance to those obtained from the active site peptides of "seryl" proteases or esterases or from other enzymes that can also be phosphorylated (alkaline phosphatase or phosphoglucomutase[72]). The seryl residue that is phosphorylated is flanked by two hydrophobic amino acids followed by an arginyl residue on the distal side. A cluster of three basic amino acids occurs nearby[25,26] and the phosphorylated site itself must belong to a highly basic region of the molecule since CNBr fragmentation of ^{32}P-labeled phosphorylase a led to the isolation of a large radioactive peptide (about 80 amino acids) with an isoelectric point of 10·5.[28] In contrast to the PLP-binding site, this peptide contains on the average one lysyl or arginyl residue for every five or six amino acids.

There are at least two reasons why we conclude that the phosphorylated site must occupy an exposed position on the surface of the protein. First, it is acted upon by both phosphorylase kinase and phosphatase which are responsible for the interconversion of phosphorylase a and b, indicating that it must also contain some of the structural features necessary for this multiple recognition. Second, it is readily attacked by

TABLE 2

Structure of the phosphorylated sites of various phosphorylases[a]

Origin	Structure	References
Rabbit skeletal muscle	C↓ T↓ T↓ (T)↓ T↓ C↓ Ser–Asp–Gln–Glu–Lys–Arg–Lys–Gln–Ile–Ser(P)–Val–Arg–Gly–Leu	26
Rabbit liver	Arg–Gln–Ile–Ser(P)–Ile–Arg	70
Human skeletal muscle	Lys–Gln–Ile–Ser(P)–Val–Arg	71
Rat skeletal muscle	Ser–Asp–Gln–Asp–Lys–Arg–Lys–Gln–Ile–Ser(P)–Val–Arg–Gly–Leu	40

[a] Amino acid residues that differ from those of the rabbit muscle phosphopeptide have been underlined.

several proteolytic enzymes: limited proteolysis by trypsin (at the arrows marked T in Table 2) removes a phosphorylated hexapeptide;[25] the remainder of the molecule which dissociates to the dimeric form (designated as phosphorylase b') is still enzymically active provided AMP is added.[25, 26, 73] Of course, it can no longer be reconverted to phosphorylase a.

As would be expected for enzymes involved in an important regulatory step, both phosphorylase kinase and phosphatase show a considerable degree of substrate specificity towards phosphorylase. Although the experiments are difficult to perform (Mg–ATP required for the kinase reaction inhibits phosphorylase phosphatase), no real evidence was obtained for a direct competition between these two enzymes for the phosphorylated site. It should be noted, however, that phosphorylase b is the substrate for phosphorylase kinase whereas phosphorylase a is the substrate for the phosphatase. No other unspecific phosphatase was found to act on phosphorylase and, conversely, phosphorylase phosphatase does not act on other phosphoproteins such as casein or phosvitin or a number of low molecular weight phosphate esters, including phosphoserine or several phosphoseryl peptides.[16] It does act, however, on the phosphorylated tetradecapeptide isolated from phosphorylase a, though rather poorly since the rate of dephosphorylation is less than 5% of that obtained with the native enzyme.[16, 25] It was of interest, therefore, to determine the minimal structure required for the reaction. Stepwise degradation of this peptide does not abolish the reaction until the arginyl group distal to the phosphoseryl residue is removed. When this occurs, the phosphatase loses all activity towards the phosphopeptide suggesting that this enzyme has a site that must be occupied by the guanido group of the substrate. This hypothesis would explain why the phosphatase is competitively inhibited by esters of lysine and arginine.[74] Contrary to what is observed when phosphorylase a is used as substrate, dephosphorylation of the phosphopeptide is neither inhibited by AMP nor accelerated by theophyllin, indicating that these compounds act by combining with phosphorylase a, not with the phosphatase.[26] This represents an interesting case in which the allosteric inhibitor affects the conformation of the substrate rather than that of the enzyme.

The dephosphorylated peptide is very slowly rephosphorylated by extracts rich in phosphorylase kinase, but it is not known whether this reaction is due to this particular enzyme, or to other non-specific protein phosphokinases. If this were the case, then kinase–kinase which contaminates most phosphorylase kinase preparations could be implicated, since it is far less demanding in terms of its substrate specificity (see Section IV F).

III. Mechanism of Action and Allosteric Properties

In spite of extensive studies of the mode of action of phosphorylase, no detailed information exists as to the exact mechanism of catalysis and the nature of the residues involved. The penetrating early investigations of the Cori's revealed the main features of the reaction. Glycogen phosphorylase catalyzes the reversible reaction:

$$\alpha\text{-D-Glucose 1-P} + \text{glycogen}_{(n)} \rightleftarrows P_i + \text{glycogen}_{(n+1)}$$

n represents the number of glucosyl residues in the polysaccharide.

Equilibrium is reached at a P_i/G-1-P ratio of 3·6 at pH 6·8, indicating that glycogen synthesis is slightly favored; since P_i is a weaker acid than G-1-P, the equilibrium is pH dependent.[1,75] In spite of this, as will be discussed in Section IV B, the enzyme functions physiologically predominantly in the direction of glycogen degradation. The phosphorylase reaction will proceed only in the presence of oligosaccharide primers of at least 3–4 glucose units. Previous reports of a *de novo* synthesis of glycogen[76] (i.e., synthesis in the absence of a primer) by purified phosphorylase could be attributed to glucose polymers contaminating the G-1-P preparations.[77] In keeping with other phosphorolytic reactions, it is the $C_{(1)}$—O bond that is cleaved[78] in both G-1-P and the $\alpha\text{-}(1 \rightarrow 4)$ glucosidic linkage.

The simplest mechanism that can be postulated is a direct nucleophilic attack by the phosphate anion on $C_{(1)}$ of the non-reducing terminal glucosyl residue of the polysaccharide. However, this mechanism is unlikely, since the phosphorylase reaction proceeds with absolute retention of configuration. A double displacement reaction is also unlikely since attempts to demonstrate the presence of a glycosyl-enzyme have failed; though, perhaps, its generation could require the presence of both substrates. No exchange was detected between P_i and G-1-P when these substrates were incubated with phosphorylase in the absence of glycogen, and no exchange between free glucose and either G-1-P or glycogen was found.[79] Similar studies provided unequivocal evidence for a double displacement reaction in sucrose phosphorylase and the glucosyl enzyme intermediate was ultimately isolated.[80] Other exchange studies have eliminated the possibility of a direct transfer of phosphate from either the PLP or the seryl phosphate residue of phosphorylase *a*. A likely mechanism would be an enzyme facilitated displacement reaction followed by frontside attack by phosphate on a stabilized carbonium ion.

Adenosine 5'-phosphate, an effector structurally unrelated to the substrates, has been known since the middle thirties to be indispensable for the activity of muscle phosphorylase b,[4, 5] and it was this very feature that led Monod to predict[81] that phosphorylase was analogous to other regulatory (allosteric) enzymes and should display cooperative kinetics. Since then this prediction has been confirmed by the combined efforts of many investigators. Cooperative kinetics can be adequately explained by either of two currently popular models. The concerted model of Monod[82] postulates an equilibrium between two conformational states of an oligomeric enzyme which differ in their affinity towards effectors and substrates. Subunit conformational changes occur in concert, i.e., symmetry is conserved; therefore, the preferential binding of a ligand to one of the states displaces the equilibrium in this direction and the resulting statistical increase in the number of binding sites generates apparent cooperative ligand binding (homotrophic interaction). When a different ligand is also capable of displacing this equilibrium, one observes heterotrophic interactions, in that one of the ligands changes the affinity of the protein for the other. The heterotrophic effect of the second ligand is normally reflected by a change in the homotrophic interactions of the first. These various effects are most pronounced when the state with the highest affinity for the ligand is present in the lowest concentration.

The sequential model of Koshland[83] generates heterotrophic interactions through ligand-induced conformational changes within a protein subunit, resulting in altered affinity for other ligands. Such effects may or may not give rise to homotrophic cooperativity depending upon the strength of the interactions between the subunits: tightly coupled subunits are capable of extensive cooperativity in that a change in conformation within one subunit will induce a similar change in the others. Several articles provide complete and excellent discussion of the details of these models.[81–85]

The two models differ significantly in several features. (a) The concerted model invokes fewer species and equilibrium reactions. (b) Homotrophic and heterotrophic interactions are necessarily linked functions in the concerted model but not in the sequential model. (c) The apparent dissociation constants assigned during cooperative ligand binding can differ only by statistical factors in the concerted model whereas the sequential model can accommodate values of any magnitude. In spite of these differences, it has proved very difficult in practice to distinguish between these alternative mechanisms and, indeed, this has been achieved in only a few instances.[86, 87]

We will not attempt to show here whether the kinetics of muscle phosphorylase are best described by one model or the other; rather the effects of the various activators and inhibitors of the enzyme will be described. It should be stated, however, that most such studies on this enzyme were carried out and interpreted with the concerted model in mind.

Table 3 lists the compounds of possible metabolic significance that are known to affect the activity of phosphorylase; their physiological role will be discussed in Section IV.

The early literature on phosphorylase contains few hints of the intriguing cooperative kinetics usually displayed by allosteric systems.

TABLE 3

Phosphorylase affectors of possible metabolic significance

Compound	Effect	K_m or $K_i{}^a$	Reference
AMP	Essential for activity of phosphorylase b	3×10^{-5} M	9
	Stimulates phosphorylase a	2×10^{-6} M	
ATP	Inhibitor of phosphorylase b	2 mM	113
	Formally competitive with AMP		
G-6-P	Inhibitor of phosphorylase b	0·3 mM	113
	Formally competitive with AMP		
UDPG	Inhibitor	0·92 mM	93
	Competitive with substrate G-1-P		
Glucose	Inhibitor of phosphorylase a		
	Competitive with G-1-P		

^{a} These values will vary considerably with the conditions of the measurements due to allosteric interactions.

However, following Monod's prediction, AMP was shown to bind to phosphorylase b in a cooperative fashion: plots of velocity against AMP concentration were clearly sigmoidal and yielded Hill coefficients of nearly two.[33] We know now that the substrates G-1-P, P_i, the activator AMP and the inhibitor ATP are all capable of homotrophic interactions. In addition, nearly all the binding sites interact with one another to produce heterotrophic effects (Table 4). For instance, the affinity of phosphorylase b for AMP is increased by two orders of magnitude (from a $K_{diss} = 5 \times 10^{-3}$ to 8×10^{-5} M) by increasing concentrations of G-1-P and P_i[88] conversely, AMP increases considerably the affinity of the enzyme for substrates and it is largely on this basis that the AMP activation of phosphorylase b is explained.[89] High concentrations of phosphate also convert the normally sigmoidal AMP binding curve to a hyperbola and the classical requirement for AMP vanishes;[90, 91] these effects are qualitatively predicted by the concerted model.

The allosteric properties of phosphorylase were probably not noticed

TABLE 4

Summary of site–site interactions in phosphorylase[a]

Sites	Catalytic	AMP	Seryl phosphate	PLP	Aggregation
Catalytic	Homo- and heterotrophic cooperativity mostly between P_i and G-1-P	Heterotrophic interaction with P_i and G-1-P	Glycogen stimulates phosphorylation (may act on kinase rather than phosphorylase)	Slight inhibition of resolution by G-1-P, P_i, UDPG	Glycogen promotes dissociation of tetramer a to dimer b
AMP	Required for b activity. Promotes active conformation. Heterotrophic cooperativity with P_i, G-1-P	Homotrophic interactions on b, none on a	Inhibits dephosphorylation	Inhibits resolution not reconstitution	Induces active conformation and aggregation to tetramers
Seryl phosphate	Freezes enzyme in active conformation. Abolishes substrate cooperativity	Increases binding by 25-fold; abolishes cooperativity	None	Inhibits resolution in the absence of glycogen	Induces active conformation and aggregation to tetramers
PLP	Indispensable for activity	Not required for binding of AMP	Not required for phosphorylation or dephosphorylation	None	Maintains quaternary structure
Aggregation	Monomers are inactive; dimer a more active than tetramer a	Binds to all forms of aggregation	No major effect on phosphorylation or dephosphorylation	Resolution greatly enhanced by monomerization	Additional aggregation site(s) become functional when primary ones are blocked

[a] This Table is meant to be read in one direction only: it summarizes the effects of the sites printed in **bold face** (left) on those in *italics* (top of the table). All cooperative effects mentioned are positive.

in early studies because the kinetics of the reaction were measured in the presence of glycerophosphate, an anion which binds to the enzyme and exerts heterotrophic effects on the binding of other ligands.[88] Indeed, sigmoidal AMP binding was first apparent only in the presence of low concentrations of the allosteric inhibitor ATP.[33] This was predicted by the concerted model which states that an inhibitor will displace the equilibrium between the two states toward the conformation with the lowest affinity for the substrate, thereby increasing the potential for cooperative interaction.

The allosteric constant L, describing the equilibrium between the active and inactive states of phosphorylase, has been determined: at 23° (measured in the absence of other ligands) it is 2100 for phosphorylase b and 3 to 13 for phosphorylase a.[88] This is in keeping with the observation that AMP is indispensable for the activity of phosphorylase b but merely stimulates phosphorylase a. Phosphorylation of the protein apparently "freezes" phosphorylase b into a form analogous to that produced by the binding of the nucleotide activator AMP. In this sense phosphorylase a and b can be looked upon as parallel cooperative systems, differing only in the details of their allosteric behavior. However, the situation is certainly not so simple since effectors such as G-6-P and ATP which are potent inhibitors of phosphorylase b have no effect on the activity of phosphorylase a.

Uridinediphosphoglucose, which acts as a competitive inhibitor towards the substrate, will actually stimulate the activity of the enzyme when present at low concentrations.[92, 93] Such an effect was initially described for aspartate transcarbamylase in the presence of succinate and was explained by assuming that binding of the competitive inhibitor to one of the subunits of the enzyme would place the other subunits in the "active" conformation.[94]

Phosphorylase, like many other regulatory proteins, can be desensitized by treatment with various agents. Limited tryptic attack[95] or reaction with sulfhydryl reagents[17] or glutaraldehyde[96] all produce desensitized enzymes which no longer display homotrophic cooperativity, yet still require AMP for activity. Specifically, AMP affects the affinity of the glutaraldehyde-treated enzyme towards other ligands. It appears as though modification of certain lysyl residues on the surface of the enzyme weakens subunit interactions to the extent that conformational changes occurring in one subunit are no longer induced in the others. As pointed out, this separation of homotrophic and heterotrophic effects is readily accommodated by the sequential model of Koshland.

The molecular architecture of phosphorylase is profoundly altered by the binding of substrates, activators and inhibitors. Initially, the striking difference between the molecular weights of phosphorylase a and b (and the fact that one is active while the other is not) made it attractive to associate activity with tetramerization. Likewise, there seemed to be a direct correlation between aggregation of the molecule and appearance of activity following the binding of allosteric effectors. This view must certainly be modified: it has been clearly shown that the dimer form of phosphorylase a is more active than the tetramer, indicating that alterations in kinetic properties may result from changes in conformation that are not necessarily accompanied by changes in quaternary structure.[97, 98] Glycogen, glucose and high salt concentrations displace the equilibrium of phosphorylase a toward its dimeric state with concomitant increase in specific activity, but the physiological significance of this activation remains unclear since phosphorylase appears to be bound to glycogen *in vivo* (see Section IV H). Similarly, AMP activates phosphorylase b and promotes dimerization at high protein concentration; yet this association does not occur under the conditions of the assay (high dilution)[31] and cannot, therefore, be obligatory for activation. Actually, one does not really know if any of these effects are of physiological significance, since the conditions under which the enzyme exists intracellularly are so vastly different from those prevailing when it is studied *in vitro*. There is no suggestion that the monomeric form of phosphorylase a or b possesses any intrinsic enzymic activity.

IV. Control of Glycogen Utilization

A. GENERAL

Several comprehensive reviews have appeared recently on glycogen metabolism[12, 99−102] and, therefore, this section will deal only with the regulation of glycogen breakdown catalyzed by phosphorylase. It is known today that the control of glycogenolysis is exercised principally at the steps catalyzed by phosphorylase, which triggers the initial formation of G-1-P from glycogen and P_i, and phosphofructokinase, which produces fructose 1,6-diphosphate from fructose 6-phosphate and ATP.[102, 103] In both instances, the reverse reactions are mediated by separate enzymes resulting in a much more effective control mechanism: metabolic flux in one direction can be blocked without impairing the flow of metabolites in the other. Also, the activities of the enzymes catalyzing the forward and reverse reactions seem to be influenced by common effectors or similar sets of reactions,[104] except that conditions leading to the activation of one shuts off the other for a "double-barrel" control at each of these steps. For instance, protein phosphorylation which results in the activation of phosphorylase and phosphorylase kinase leads to the inactivation (I to D conversion) of glycogen synthetase.[105] Likewise, a high ATP/AMP ratio activates fructose-1,6-diphosphatase and inhibits phosphofructokinase—a major cause for the decrease of glycolysis by respiration known as the Pasteur Effect.[106] In fact, Passonneau and Lowry showed that when glycolysis in muscle is triggered by electric stimulation, both phosphorylase and phosphofructokinase increase their activity synchronously, thus indicating an even higher level of sophistication in the regulation of the overall pathway.[102]

B. MOLECULAR CONTROL OF PHOSPHORYLASE ACTIVITY

While the reaction catalyzed by glycogen phosphorylase is fully reversible ($K = 3\cdot6$ for the formation of glycogen at neutrality), several observations support the view that, under physiological conditions, this enzyme functions almost entirely in the direction of glycogen breakdown:[107] (a) glycogen synthesis occurs even at high P_i/G-1-P ratios that favor glycogen breakdown, indicating that the synthesis must occur through a separate pathway; (b) physiological conditions leading to an increase in phosphorylase activity, such as the administration of epinephrine or glucagon, always bring about a breakdown of glycogen; (c) patients suffering from McArdle's disease, a hereditary myopathy in which phosphorylase is lacking,[108, 109] or I-strain mice deficient in phosphorylase kinase,[110] generally show an accumulation of glycogen.

Numerous reports indicate that in resting muscle, phosphorylase exists predominantly in the inactive b form.[111] Activation of phosphorylase can occur either through non-covalent changes in conformation brought about by effectors such as AMP or through covalent modification of the protein. It has been a very difficult task to determine the relative contribution of these two separate mechanisms to the state of activity of the enzyme.

C. CONTROL OF PHOSPHORYLASE ACTIVITY BY AMP

As indicated earlier, purified phosphorylase b is activated by AMP at concentrations which vary with the conditions of the reaction, including the nature and concentration of substrates, divalent metal ions, pH, temperature, etc. (see Section III).

In spite of the fact that AMP is rapidly utilized in muscle (either by conversion to IMP or rephosphorylation to ATP via ADP), the measured level of this nucleotide is sufficiently high to suggest that phosphorylase b should always be active.[112] Yet this activity is not observed, probably because of the relatively high levels of the inhibitors ATP (which competes for the AMP site) and G-6-P (which seems to bind at a separate site).[113]

In addition, changes in the level of nucleotide effectors in the muscle are not fast enough to account for the rapid (10–20 s) activation of phosphorylase observed upon electric stimulation.[102] Conclusions of this kind are based, of course, on the assumptions that within the muscle cell (a) there is no compartmentation where AMP could preferentially accumulate; and (b) phosphorylase would behave as a solution of pure enzyme. The latter is a dangerous assumption, since muscle phosphorylase, together with several other enzymes involved in glycogen metabolism, appears to be associated with a glycogen particulate fraction (see Section IV H).

Even though the AMP activation of phosphorylase b appears to play a minor role as compared to the covalent conversion of phosphorylase b to a, it must be physiologically operative. I-strain mice which lack phosphorylase kinase due to a genetic defect in their X-chromosome[114] (and, therefore, cannot form phosphorylase a from b) nonetheless produce G-6-P and lactate from glycogen; presumably this occurs by means of AMP activation of the enzyme.[115]

D. COVALENT ACTIVATION OF PHOSPHORYLASE AND FORMATION OF PHOSPHO–DEPHOSPHO HYBRIDS

Conversion of phosphorylase b to a requires phosphorylase kinase, Mg^{2+}, and ATP; the reverse reaction is catalyzed by phosphorylase phosphatase. When both enzymes act simultaneously, the overall reaction is that of an ATPase:

$$\text{Phosphorylase } b + 4\text{ATP} \xrightarrow{\text{kinase}} \text{phosphorylase } a + 4\text{ADP}$$

$$\text{Phosphorylase } a + 4\text{H}_2\text{O} \xrightarrow{\text{phosphatase}} \text{phosphorylase } b + 4\text{P}_i$$

$$\text{SUM:}\quad 4\text{ATP} + 4\text{H}_2\text{O} \longrightarrow 4\text{ADP} + 4\text{P}_i$$

Apparently, however, phosphorylation or dephosphorylation of phosphorylase does not proceed in an all-or-none fashion in which, at any one time, one would find only a mixture of the fully phosphorylated a form and the fully dephosphorylated b form.[116] On the contrary, we have proposed that these reactions proceed in a step-wise manner in which partially phosphorylated intermediates are produced according to the scheme depicted in Fig. 8 for the $a \to b$ reaction.[117]

This assumption was based on the following set of observations: The $a \to b$ conversion can be followed either by disappearance of phosphorylase a activity (measured in the absence of AMP) or by loss of bound [32]P. If the two reactions are compared, essentially identical rates are observed provided enzymic activity is measured at low concentrations of G-1-P (K_m[G-1-P] for phosphorylase a is 0·016 M). However, at high concentrations of G-1-P ($> 0·075$ M), little loss of activity is observed until approximately half the protein-bound phosphate has been hydrolyzed. Furthermore, addition of as little as 10^{-3} M G-6-P to the assay totally suppresses this abnormally high initial activity, lowering it to a level below that of the bound phosphate. There is no effect of

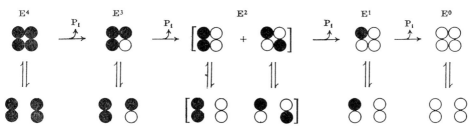

FIG. 8. Model for the formation of phospho–dephospho hybrids during the phosphorylase a to b conversion.

G-6-P on the activity of phosphorylase a itself, as previously demonstrated.[113]

We therefore advanced the following hypothesis.[117] Phosphorylase may exist in two conformations (Fig. 9): active (squares) and inactive (circles) ones, in equilibrium with one another. In phosphorylase b this equilibrium strongly favors the inactive conformation unless positive

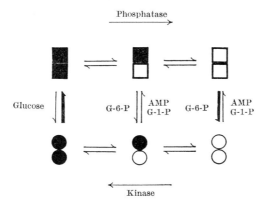

FIG. 9. Model for the structure and conformation of phospho-dephospho hybrids. Circles represent inactive conformations, and squares, active ones. Open and filled symbols represent non-phosphorylated and phosphorylated subunits, respectively.

effectors such as AMP or high concentrations of P_i are added; G-6-P, on the other hand, favors the inactive conformation. By contrast, equilibrium for the phosphorylated form (black in Fig. 9) of the protein (phosphorylase a) is strongly displaced towards the active conformation to the extent that the inactive, phosphorylated dimer is only seen under special conditions. For instance, incubation of phosphorylase a with glucose promotes the formation of an inactive dimer than can be converted to the active state by positive effectors such as AMP or substrates.[118] The active conformation has a strong tendency to aggregate and this is why phosphorylase a (or phosphorylase b in the presence of AMP) is usually found in the tetrameric state.

Unlike fully phosphorylated phosphorylase a, or fully dephosphorylated b, the equilibrium between the active and inactive conformations of the partially phosphorylated hybrids would be much more sensitive towards positive or negative effectors, i.e., much more easily shifted from one state to the other. Perhaps, this simply results from the fact that phosphorylation of only one of the subunits introduces an element of dissymmetry in the molecule that prevents it from settling down in the conformations of the parent molecules. In the presence of high concentrations of G-1-P, the material would be fully active; on the other hand, G-6-P at concentrations of 10^{-3} M or below would displace the equilibrium entirely towards the inactive state.

These effects can also be readily visualized in the ultracentrifuge since the active conformation tends to aggregate to a tetrameric state whereas the inactive conformation remains as a dimer. Addition of G-1-P to a mixture of "phospho-dephospho" hybrids yield predominantly tetramers while addition of G-6-P will produce mostly dimers. Such an effect is much less pronounced when a mixture of pure phosphorylase b and a is treated in similar fashion.

A partial separation of the hybrids was obtained by TEAE-cellulose chromatography. Unfortunately, no pure hybrid preparation could be obtained since the heterologous species rapidly dismutate to the thermodynamically more stable homologous forms of phosphorylase a and b.

Both Mg^{2+} and ATP are plentiful in resting muscle (approximately 10 mM and 7 mM, respectively)[102] and yet, phosphorylase remains in its inactive b form because phosphorylase kinase itself is essentially inactive at physiological pH (about 6·8); kinase activity is expressed, however, if the pH is raised to about 8·2–8·6.[119–121] Breakdown of glycogen will be triggered by the conversion of nonactivated phosphorylase kinase to that form which is active at neutrality. Apparently, this latter reaction can proceed by at least two mechanisms, the interrelations of which are not yet fully understood, namely (a) addition of Ca^{2+} ions, and

(b) addition of Mg-ATP, this latter reaction being accelerated by cyclic-3',5'-AMP. The reactions are summarized in Fig. 10.

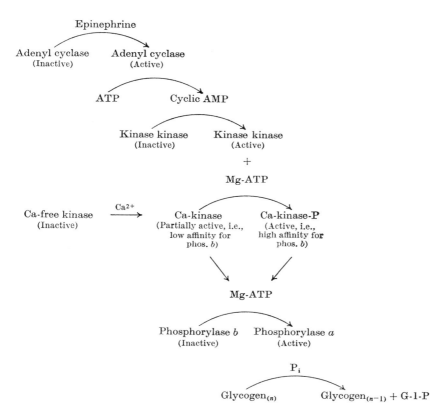

FIG. 10. Schematic representation of the activation of rabbit muscle phosphorylase. For the sake of simplicity, the reverse processes have not been included.

E. ACTIVATION OF PHOSPHORYLASE KINASE BY Ca^{2+}

Phosphorylase kinase has been obtained in homogeneous form.[121] It is a large protein of mol.wt. $1\cdot27 \times 10^6$ apparently made up of approximately 24 subunits of mol.wt. about 50,000; it is not known whether or not the subunits are identical (E. G. Krebs, personal communication).

When phosphorylase kinase was first observed in crude muscle extracts, this enzyme was found to be essentially inactive at physiological pH, but could be activated by addition of various divalent metal ions, including Ca^{2+}.[122] Subsequently, it was determined that a purified fraction of phosphorylase kinase was no longer activated by calcium unless a protein fraction that was removed during the enzyme purification was re-added.[123] This protein, named kinase activating factor or KAF, was later found to be a calcium-requiring proteolytic enzyme that brought about a transient activation of phosphorylase kinase through partial proteolysis.[124] Similar activation by other proteolytic enzymes such as trypsin and chymotrypsin had been previously noted. The irreversibility of this hydrolytic process appeared to eliminate the possibility that KAF was involved in a physiological control mechanism.

Recently, however, new evidence was obtained indicating that, in addition to the KAF-mediated reaction, a reversible Ca^{2+} activation of kinase existed that would certainly be of physiological significance.[125, 126] It was observed that kinase partially inactivated by chelating agents such as EDTA or EGTA could be reactivated by the addition of Ca^{2+}, Sr^{2+}, or Mn^{2+}, suggesting that these metal ions were directly involved in kinase activity. This assumption was given strong support by the finding that Ca^{2+}-free kinase (obtained by EGTA treatment) was essentially inactive, but could be reactivated by concentrations of Ca^{2+} of the order of 10^{-7} M.[125]

Calcium activation (as well as the activation by Mg–ATP described in the next section) seems to result from a higher affinity of the enzyme for its substrate, phosphorylase b. It will, therefore, become evident predominantly at low phosphorylase b concentrations and not under saturating conditions.

The reversible activation of kinase by calcium, unlike the irreversible KAF-activation described above, supports the view that this enzyme behaves as a Ca^{2+}-requiring metalloenzyme, and that the overall process is intimately linked to that of muscle contraction. It is now generally accepted[127] that muscle contraction is triggered by the release of Ca^{2+} ions from the sarcoplasmic reticulum, particularly their terminal cisternae, following depolarization of the T-tubules (transverse tubules or intermediary vesicles) as a result of nerve impulse. Calcium released in the myoplasm is believed to interact with troponin, a globular protein of mol.wt. approximately 50,000 which binds close to 2 g-atom Ca^{2+}/mol with a binding constant of $1\cdot3 \times 10^{-6}$ M; in doing so, it releases an inhibitory effect displayed by Ca-free troponin on the interaction of actin and myosin. This inhibitory effect is not direct, but mediated through the presence of tropomyosin to which troponin is bound. Following contraction, calcium ions are reabsorbed by the sarcoplasmic reticulum in an ATP-dependent reaction, thus restoring the muscle to its initial resting state.

TABLE 5

Comparative properties of phosphorylases

Enzyme source	PLP as cofactor	Molecular weight (dimeric form)	Activation following phosphorylation	Tetramerization following phosphorylation	Requirement for AMP	Reference
Rabbit muscle (Isoenzyme III)	+	185,000	+	+	+	144, 145
Rabbit heart (Isoenzyme I)	+	200,000	+	−	+	146, 70
Rabbit liver[a]	+	185,000	+	−	−	24, 147
Human muscle	+	177,000	+	+	+	148
Human platelet	+	185,000	+	+	+	40
Rat muscle	+	185,000	+	+	+	149
Frog muscle	+	188,000	+	−	+	150
Dogfish muscle	+	200,000	+	−	+	151, 152
Lobster tail muscle	+	180,000	+	−	+	153
Yeast	+	200,000	+	−[b]	−	154
Neurospora	+	320,000	−	−	+[c]	155, 156
E. coli	+	250,000	−	−	+[c]	157, 158
Potato	+	207,000	−	−	−	157, 158

[a] Dephosphorylated form inactive in the presence of AMP.
[b] Tetramer form (inactive?) also seen.
[c] Slight activation.

3

variations; (c) all enzymes from animal and yeast appear to be activated by phosphorylation of the protein but not those from plants and certain protista (*Neurospora* and *E. coli*); (d) considerable differences exist concerning tetramerization of the phosphorylated form of the enzyme: whereas phosphorylases from rabbit, human, rat and frog muscle (as well as that from human platelets) tetramerize readily, all others including those from rabbit heart (isoenzyme I) and liver do not. Thus, tetramerization does not appear to be an essential requirement for catalysis, a conclusion reached previously from the study of rabbit muscle phosphorylase a;[97, 98] (e) likewise, not all enzymes are activated by AMP: again, most animal phosphorylases seem to rely on this form of activation, in contrast to the protista and plant enzymes; (f) all phosphorylases display an optimum pH of activity about 6·5 (5·8 for the yeast enzyme) and most have turnover numbers of the order of 8000 μmol substrate converted/min per μmol enzyme protomer.

Rabbit heart muscle has been shown to possess three chromatographically or electrophoretically distinct types of phosphorylase designated as I, II and III.[144, 145] All three undergo activation by phosphorylase kinase to yield at least six different electrophoretic species. The b forms of heart phosphorylase (I^b, II^b and III^b) are analogous to skeletal muscle phosphorylase b in that they are activated by AMP, and the a forms are active in the absence of this effector. Isoenzyme III was found to be identical with the skeletal muscle enzyme, in contrast to isoenzyme I which does not tetramerize after phosphorylation. Isoenzyme II^b is a hybrid of isoenzymes I and III. Patients suffering from McArdle's disease and who lack skeletal muscle phosphorylase[108, 109] are presumably deficient in heart isoenzyme III; since type I is genetically distinct, its presence in this tissue should not be affected.

Hybridization undoubtedly proceeds through monomerization of the oligomeric species and, therefore, is favored by conditions promoting dissociation such as high ionic strength, imidazole citrate buffer, high temperature, etc. Adenosine 5'-phosphate and G-6-P, which oppose monomerization, inhibit this process.

Liver phosphorylase represents a different type of enzyme, in that its dephosphorylated form is inactive even in the presence of AMP;[159] activation results only from phosphorylation of the protein. Inactive liver phosphorylase is analogous to heart isoenzyme I, in that both its phosphorylated and dephosphorylated forms have identical sedimentation constants; its electrophoretic mobility on the other hand is similar to that of isoenzyme III. Liver phosphorylase does not hybridize with either isoenzyme I or III, underlining the gross structural differences between these various molecular species.[145] When the overall amino acid composition of rabbit liver phosphorylase [70] was compared with that of known muscle phosphorylases in terms of their compositional divergence factor,[160] a greater difference was observed between rabbit liver and rabbit muscle phosphorylase than between the muscle enzymes from rabbit, rat, man, frog and dogfish.[150] Furthermore, liver and skeletal muscle enzymes show no immunological cross reactivity,[161] even though the amino acid sequences about their phosphorylated sites are very similar.

Human platelet phosphorylase[148] has a molecular weight, subunit structure, allosteric and covalent activation similar to that of the muscle enzyme. The equilibrium between tetrameric and dimeric a forms, however, is faster than with the muscle enzyme. Agents that cause agglutination also promote glycogenolysis; yet, no change in the ratio of phosphorylase b to a was observed. It was proposed that in this instance, intracellular changes of substrates and modifiers brought about the activation of the enzyme.

Yeast is able to accumulate glycogen up to 50% of its dry weight when grown under anaerobic conditions or on a high-carbohydrate-/low-nitrogen-containing medium.[162] Contact with oxygen or transfer into a nitrogen-rich medium initiates an immediate breakdown of this reserve polysaccharide. Glycogen also accumulates in certain bacteria during the stationary phase, perhaps as a result of an excess production of ATP. Both yeast[153] and an inducible $E. coli$ phosphorylase[156] have been obtained in homogeneous form. The former enzyme appears to be phosphorylated by a yeast phosphorylase kinase (it is not affected by muscle kinase), with concomitant activation. However, the activated form of the enzyme is in the dimeric state: a tetramer form was also observed, but it appears to be essentially inactive. No information is at present available as to the structure of the phosphorylated site. In contrast, no phosphorylation was detected for the $E. coli$ enzyme; this organism also produces a constitutive phosphorylase.

Potato phosphorylase[158] is the only specimen of this enzyme that has been obtained in good yield and purity from higher plants. It has a higher affinity toward starch than glycogen and, like its animal counterpart, does not appear to promote the $de\ novo$ synthesis of polysaccharides. Potato phosphorylase has no phosphorylated site and is not stimulated by AMP; it lacks the characteristics of an associating–dissociating system and, perhaps for this reason, shows greater stability towards high concentrations of urea or deforming agents than other phosphorylases. Sodium borohydride reduction results in an enzyme which retains 50% of its original activity. Surprisingly, the reduced enzyme lacks the 330 nm absorption peak, characteristic of reduced muscle phosphorylase. This might indicate some differences in the structure of the cofactor.

VI. Concluding Remarks

Although much information has accumulated on the chemical and physical properties of phosphorylase, it is obvious from this review that information concerning many crucial aspects of this enzyme is still sorely lacking.

From Table 5, it can be seen that all glycogen phosphorylases consist of a basic subunit of very similar size (mol.wt. approximately 100,000), and contain one equivalent of PLP. It appears reasonable to assume, therefore, that these proteins all originated through the same ancestral gene which, perhaps, also gave rise to glycogen synthetase; these two enzymes have many properties in common including similar amino acid sequences at their phosphorylated sites. Why do we need such a large basic subunit? At first glance, one could speculate that a subunit of considerable size might be required in order to accommodate so many different binding sites, i.e., two control sites, a cofactor-binding site, perhaps a separate G-6-P binding site, in addition to a catalytic site capable of binding three different substrates. But then, why should potato phosphorylase be of similar size, since it appears to display neither a phosphorylated nor an AMP-binding site?

The enzymes involved in the phosphorylation and dephosphorylation of the protein have been fairly well characterized. On the other hand, we have no information as to how phosphorylation of the protein brings about a change in conformation that results in a loss of cooperative interactions and appearance of enzymic activity. We do not know whether those phosphorylases that are not regulated by a phosphorylation–dephosphorylation mechanism lack the proper sequence at this site or if this segment of the molecule is present but inaccessible to phosphorylating enzymes.

Even less is known about the function of PLP in phosphorylase. This is unfortunate because as indicated above, all glycogen phosphorylases contain this cofactor and information regarding the comparative sequences, spectral characterization and enzymic properties of this site might be of considerable significance from a mechanistic and evolutionary point of view.

It is worthwhile to emphasize once again the role played by adenine nucleotides in the regulation of enzymic activity.[104] Since they represent one of the main forms of energy storage available for cellular processes, their utilization as feedback regulators is of obvious advantage in the control of certain pathways, particularly those concerned with energy metabolism. Of course, adenine nucleotides would be expected to bind readily to proteins by virtue of their hydrophilic character, multiple charges and polycyclic nature. It would be interesting to know, therefore, if the many enzymes that are controlled by interactions with this class of compounds have structural features in common.

Essentially nothing is known about the constellation of groups forming the catalytic site of the enzyme and their spatial arrangement. There is overwhelming evidence that carboxyl side chains are directly involved in the mechanism of action of both lysozyme[163] and sucrose phosphorylase,[80] two enzymes catalyzing the cleavage of glycosidic bonds. Perhaps glycogen phosphorylase functions by a similar mechanism, although as yet there is no evidence for or against this possibility.

Several of the questions raised above will be clarified if not answered when the primary and tertiary structures of the enzyme are resolved by the techniques of sequence analysis and X-ray crystallography. There will remain, however, the very complex problem as to how phosphorylase activity and glycogen metabolism in general are regulated in the intact organism.

The study of the phosphorylase system has opened the way to our comprehension of the molecular basis of hormonal action.[164, 165] Remarkable advances have been made lately in this area as well as in the mechanism of muscle contraction.[108, 127, 166, 167] Hopefully, we are approaching the time when the complex interrelationships among these important physiological processes will be fully understood.

REFERENCES

1. Cori, C. F., Cori, G. T. & Green, A. A. (1943). Crystalline muscle phosphorylase III. Kinetics. *J. biol. Chem.* **151**, 39–55.
2. Krebs, E. G. & Fischer, E. H. (1956). The phosphorylase *b* to *a* converting enzyme of rabbit skeletal muscle. *Biochim. biophys. Acta,* **20**, 150–157.
3. Madsen, N. B. & Cori, C. F. (1956). The interaction of muscle phosphorylase with *p*-chloromercuribenzoate I. Inhibition of activity and effect on the molecular weight. *J. biol. Chem.* **223**, 1055–1065.
4. Cori, C. F. & Cori, G. T. (1936). Mechanisms of formation of hexose-monophosphate in muscle and isolation of a new phosphate ester. *Proc. Soc. exp. Biol. Med.* **34**, 702–705.
5. Parnas, M. J.-K. & Mochnacka, I. (1936). Le role de l'acide inosique dans la glycogenolyse musculaire. *C. r. Soc. Biol.* **123**, 1173–1175.
6. Baranowski, T., Illingworth, B., Brown, D. H. & Cori, C. F. (1957). The isolation of pyridoxal 5'-phosphate from crystalline muscle phosphorylase. *Biochim. biophys. Acta,* **25**, 16–21.
7. Fischer, E. H., Forrey, A. W., Hedrick, J. L., Hughes, R. C., Kent, A. B.

& Krebs, E. G. (1963). Pyridoxal 5'-phosphate in the structure and function of phosphorylase. In *Chemical and Biological Aspects of Pyridoxal Catalysis*, pp. 543–562. Ed. by Snell, E. E., Fasella, P. M., Braunstein, A. & Rossi-Fanelli, A. Oxford: Pergamon Press.

8. Miller, S. L. & Urey, H. C. (1959). Organic compound synthesis on the primitive Earth. *Science*, N.Y. **130**, 245–251.

9. Brown, D. H. & Cori, C. F. (1960). Animal and plant polysaccharide phosphorylase. In *The Enzymes*, 2nd ed., vol. 5, 207–228. Ed. by Boyer, P. D., Lardy, H. & Myrbäck, K. New York: Academic Press.

10. Krebs, E. G. & Fischer, E. H. (1962). Molecular properties and transformations of glycogen phosphorylase in animal tissues. *Adv. Enzymol.* **24**, 263–290.

11. Fischer, E. H. & Krebs, E. G. (1966). Relationship of structure to function of muscle phosphorylase. *Fedn Proc. Fedn Am. Socs. exp. Biol.* **25**, 1511–1520.

12. Helmreich, E. (1969). Control of synthesis and breakdown of glycogen, starch and cellulose. In *Comprehensive Biochemistry*, vol. 17, pp. 17–92. Ed. by Florkin, M. & Stotz, E. H. Amsterdam: Elsevier.

13. Cori, G. T. & Cori, C. F. (1945). The enzymatic conversion of phosphorylase *a* to *b*. *J. biol. Chem.* **158**, 321–332.

14. Keller, P. J. & Cori, G. T. (1953). Enzymatic conversion of phosphorylase *a* to phosphorylase *b*. *Biochim. biophys. Acta*, **12**, 235–238.

15. Keller, P. J. & Cori, G. T. (1955). Purification and properties of the phosphorylase-rupturing enzyme. *J. biol. Chem.* **214**, 127–134.

16. Graves, D. J., Fischer, E. H. & Krebs, E. G. (1960). Specificity studies on muscle phosphorylase phosphatase. *J. biol. Chem.* **235**, 805–809.

17. Kastenschmidt, L. L., Kastenschmidt, J. & Helmreich, E. (1968). Subunit interactions and their relationship to the allosteric properties of rabbit skeletal muscle phosphorylase *b*. *Biochemistry, Easton*, **7**, 3590–3608.

18. Shaltiel, S., Hedrick, J. L. & Fischer, E. H. (1966). On the role of pyridoxal 5'-phosphate in phosphorylase II. Resolution of rabbit muscle phosphorylase *b*. *Biochemistry, Easton*, **5**, 2108–2116.

19. Seery, V. L., Fischer, E. H. & Teller, D. C. (1967). A reinvestigation of the molecular weight of glycogen phosphorylase. *Biochemistry, Easton*, **6**, 3315–3327.

20. DeVincenzi, D. L. & Hedrick, J. L. (1967). Re-evaluation of the molecular weights of glycogen phosphorylase *a* and *b* by using Sephadex gel filtration. *Biochemistry, Easton*, **6**, 3489–3497.

21. Ullmann, A., Goldberg, M. E., Perrin, D. & Monod, J. (1968). On the determination of molecular weight of proteins and protein subunits in the presence of 6 M guanidine hydrochloride. *Biochemistry, Easton*, **7**, 261–265.

22. Weber, K. & Osborn, M. (1969). The reliability of molecular weight determinations by dodecylsulfate–polyacrylamide gel electrophoresis. *J. biol. Chem.* **244**, 4406–4412.

23. Seery, V. L., Fischer, E. H. & Teller, D. C. (1970). The subunit structure of glycogen phosphorylase. *Biochemistry, Easton*, in press.

24. Appleman, M. M., Yunis, A. A., Krebs, E. G. & Fischer, E. H. (1963). Comparative studies on glycogen phosphorylase. V. The amino acid composition of rabbit and human skeletal muscle phosphorylase. *J. biol. Chem.* **238**, 1358–1361.

25. Fischer, E. H., Graves, D. J., Crittenden, E. R. & Krebs, E. G. (1959).

Structure of a site phosphorylated in the phosphorylase b to a reaction. *J. biol. Chem.* **234**, 1698–1704.

26. Nolan, C., Novoa, W. B., Krebs, E. G. & Fischer, E. H. (1964). Further studies on the site phosphorylated in the phosphorylase b to a reaction. *Biochemistry, Easton,* **3**, 542–551.

27. Zarkadas, C. G., Smillie, L. B. & Madsen, N. B. (1968). SH groups of muscle phosphorylase II. Thiol sequences and subunit structure. *J. molec. Biol.* **38**, 245–247.

28. Saari, J. C., unpublished results.

29. Valentine, R. C. & Chignell, D. A. (1968). Electron microscopy of muscle phosphorylase molecules: Significance of a rhombic shape. *Nature, Lond.* **218**, 950–953.

30. Chignell, D. A., Gratzer, W. B. & Valentine, R. C. (1968). Subunit interaction in native and modified muscle phosphorylases. *Biochemistry, Easton,* **7**, 1082–1089.

31. Ullmann, A., Vagelos, P. R. & Monod, J. (1964). The effect of 5′-adenylic acid upon the association between bromthymol blue and muscle phosphorylase b. *Biochem. biophys. Res. Commun.* **17**, 86–92.

32. Parmeggiani, A. & Morgan, H. E. (1962). Effect of adenine nucleotides and inorganic phosphate on muscle phosphorylase activity. *Biochem. biophys. Res. Commun.* **9**, 252–256.

33. Madsen, N. B. (1964). Allosteric properties of phosphorylase b. *Biochem. biophys. Res. Commun.* **15**, 390–395.

34. Harris, J. I. & Perham, R. N. (1965). Glyceraldehyde 3-phosphate dehydrogenases I. The protein chains in glyceraldehyde 3-phosphate dehydrogenase from pig muscle. *J. molec. Biol.* **13**, 876–884.

35. Sevilla, C. L. (1969). The purification and characterization of rat skeletal muscle glycogen phosphorylase. Ph.D. Thesis, University of Washington.

36. Moav, B. & Harris, T. N. (1967). Pyrrolid-2-one-5 carboxylic acid involvement in the biosynthesis of rabbit immunoglobulin. *Biochem. biophys. Res. Commun.* **29**, 773–776.

37. Ambler, R. P. (1967). Enzymic hydrolysis with carboxypeptidases. In *Methods in Enzymology,* vol. 11, pp. 155–166. Ed. by Hirs, C. H. W. New York: Academic Press.

38. Braun, V. & Schroeder, W. A. (1967). A reinvestigation of the hydrazinolytic procedure for the determination of C-terminal amino acids. *Archs Biochem. Biophys.* **118**, 241–252.

39. Matsuo, H., Fujimoto, Y. & Tatsuno, T. (1966). A novel method for the determination of C-terminal amino acid in polypeptides by selective tritium labeling. *Biochem. biophys. Res. Commun.* **22**, 69–74.

40. Sevilla, C. L. & Fischer, E. H. (1969). The purification and properties of rat muscle phosphorylase. *Biochemistry, Easton,* **8**, 2161–2171.

41. DuVigneaud, V. (1954–55). Hormones of the posterior pituitary gland: Oxytocin and Vasopressin. *Harvey Lect. Ser.* **50**, pp. 1–26.

42. Fischer, E. H., Kent, A. B., Sneider, E. R. and Krebs, E. G. (1958). The reaction of sodium borohydride with muscle phosphorylase. *J. Am. chem. Soc.* **80**, 2906.

43. Madsen, N. B. (1956). The interaction of muscle phosphorylase with p-chloromercuribenzoate II. A study by light scattering. *J. biol. Chem.* **223**, 1067–1076.

44. Madsen, N. B. & Gurd, F. R. N. (1956). The interaction of muscle phos-

phorylase with p-chloromercuribenzoate III. The reversible dissociation of phosphorylase. *J. biol. Chem.* **223**, 1075–1087.

45. Damjanovich, S. & Kleppe, K. (1966). The reactivity of SH groups in phosphorylase *b*. *Biochim. biophys. Acta*, **122**, 145–147.

46. Gold, A. M. (1968). Sulfhydryl groups of rabbit muscle glycogen phosphorylase *b*. Reaction with dinitrophenylating agents. *Biochemistry, Easton*, **7**, 2106–2115.

47. Battell, M. L., Zarkadas, C. G., Smillie, L. B. & Madsen, N. B. (1968). The SH groups of muscle phosphorylase III. Identification of cysteinyl peptides related to function. *J. biol. Chem.* **243**, 6202–6209.

48. Avramovic-Zikic, O., Smillie, L. B. & Madsen, N. B. (1970). The sulfhydryl groups of muscle phosphorylase IV. Reactivities as related to changes in protein structure. *J. biol. Chem.*, in press.

49. Battell, M. L., Smillie, L. B. & Madsen, N. B. (1968). The SH groups of muscle phosphorylase I. Number and reactivity. *Can. J. Biochem.* **46**, 609–615.

50. Snell, E. E. (1958). Chemical structure in relation to biological activities of vitamin B_6. *Vitam. Horm.* **16**, 77–125.

51. Guirard, B. M. & Snell, E. E. (1964). Vitamin B_6 function in transamination and decarboxylation reactions. In *Comprehensive Biochemistry*, vol. 15, pp. 138–199. Ed. by Florkin, M. and Stotz, E. H. Amsterdam: Elsevier.

52. Illingworth, B., Jansz, H. S., Brown, T. H. & Cori, C. F. (1958). Observations on the function of pyridoxal 5'-phosphate in phosphorylase. *Proc. natn. Acad. Sci. U.S.A.* **44**, 1180–1191.

53. Fischer, E. H. (1964). Binding of vitamin B_6-coenzymes and labelling of active site of enzymes by sodium borohydride reduction. In *Structure and Activity of Enzymes*, pp. 111–120. Ed. by Goodwin, T. W., Harris, J. I. & Hartley, B. S. Sympos. No. 1, Fedn European Biochemical Societies. London: Academic Press.

54. Hedrick, J. L. & Fischer, E. H. (1965). On the role of pyridoxal 5'-phosphate in phosphorylase I. Absence of classical vitamin B_6-dependent enzymatic activities in muscle glycogen phosphorylase. *Biochemistry, Easton*, **4**, 1337–1343.

55. Hedrick, J. L., Shaltiel, S. & Fischer, E. H. (1969). Conformational changes and the mechanism of resolution of glycogen phosphorylase *b*. *Biochemistry, Easton*, **8**, 2422–2429.

56. Fischer, E. H. & Krebs, E. G. (1964). Phosphorylase and related enzymes of glycogen metabolism. *Vitam. Horm.* **22**, 399–410.

57. Forrey, A. W., Sevilla, C. L., Saari, J. C. & Fischer, E. H., manuscript in preparation.

58. Takahashi, K. (1965). The amino acid sequence of ribonuclease T_1. *J. biol. Chem.* **240**, PC 4117–4119.

59. Anderson, F. J. & Martell, A. E. (1964). Pyridoxal phosphate: Molecular species in solution. *J. Am. chem. Soc.* **86**, 715–720.

60. Metzler, D. E. & Snell, E. E. (1955). Spectra and ionization constants of the vitamin B_6 group and related 3-hydroxypyridine derivatives. *J. Am. chem. Soc.* **77**, 2431–2437.

61. Bruice, T. C. & Topping, R. M. (1963). Catalytic reactions involving azomethines (Papers I, II & III). *J. Am. chem. Soc.* **85**, 1480–1496.

62. Kent, A. B., Krebs, E. G. & Fischer, E. H. (1958). Properties of crystalline phosphorylase *b*. *J. biol. Chem.* **232**, 549–558.

63. Shaltiel, S., Hedrick, J. L., Pocker, A. & Fischer, E. H. (1969). Reconstitution of apophosphorylase with pyridoxal 5'-phosphate analogs. *Biochemistry, Easton*, **8**, 5189–5196.

64. Shaltiel, S., Hedrick, J. L. & Fischer, E. H. (1969). Stereospecific requirements for carbonyl reagents in the resolution and reconstitution of phosphorylase *b*. *Biochemistry, Easton*, **8**, 2429–2436.

65. Hedrick, J. L., Shaltiel, S. & Fischer, E. H. (1966). On the role of pyridoxal 5'-phosphate in phosphorylase III. Physicochemical properties and reconstitution of apophosphorylase *b*. *Biochemistry, Easton*, **5**, 2117–2125.

66. Hullar, T. L. (1967). Vinyl phosphonates: A convenient route to phosphonic acid analogues of phosphate monoesters. *Tetrahedron Lett.* **49**, 4921–4923.

67. Pocker, A. & Fischer, E. H. (1969). Synthesis of analogs of pyridoxal 5'-phosphate. *Biochemistry, Easton*, **8**, 5181–5188.

68. Vidgoff, J., unpublished results.

69. Bresler, S. & Firsov, L. (1968). Spectrophotometric study of the enzyme-substrate complex of phosphorylase *b*. *J. molec. Biol.* **35**, 131–141.

70. Wolf, D. P., Fischer, E. H. & Krebs, E. G. (1970). Amino acid sequence of the phosphorylated site in rabbit liver glycogen phosphorylase. *Biochemistry, Easton*, in press.

71. Hughes, R. C., Yunis, A. A., Krebs, E. G. & Fischer, E. H. (1962). Comparative studies on glycogen phosphorylase III. The phosphorylated site in human muscle phosphorylase *a*. *J. biol. Chem.*, **237**, 40–43.

72. Winter, W. D. & Neurath, H. (1968). Enzyme active sites. In *Handbook of Biochemistry*, K16–18. Ed. by Sober, H. A.

73. Keller, P. J. (1955). The action of trypsin on phosphorylase *a*. *J. biol. Chem.* **214**, 135–141.

74. Keller, P. J. & Fried, M. (1955). Inhibition of the phosphorylase rupturing enzyme by some trypsin substrates. *J. biol. Chem.* **214**, 143–148.

75. Cori, G. T. & Cori, C. F. (1940). The kinetics of the enzymatic synthesis of glycogen from glucose-1-phosphate. *J. biol. Chem.* **135**, 733–756.

76. Cori, G. T. & Cori, C. F. (1943). Crystalline muscle phosphorylase IV. Formation of glycogen. *J. biol. Chem.* **151**, 57–63.

77. Abdullah, M., Fischer, E. H., Qureshi, M. Y., Slessor, K. N. & Whelan, W. J. (1965). Requirement of rabbit muscle glycogen phosphorylase for primer. *Biochem. J.* **97**, 9 P.

78. Cohn, M. (1949). Mechanisms of cleavage of glucose-1-phosphate. *J. biol. Chem.* **180**, 771–781.

79. Cohn, M. & Cori, G. T. (1948). On the mechanism of action of muscle and potato phosphorylase. *J. biol. Chem.* **175**, 89–93.

80. Voet, J. G. & Abeles, R. H. (1970). The mechanism of action of sucrose phosphorylase. *J. biol. Chem.* **245**, 1020–1031.

81. Monod, J., Changeux, J.-P., Jacob, F. (1963). Allosteric proteins and cellular control systems. *J. molec. Biol.* **6**, 306–329.

82. Monod, J. Wyman, J., Changeux, J.-P. (1965). On the nature of allosteric transition: A plausible model. *J. molec. Biol.* **12**, 88–118.

83. Koshland, D. E., Nemethy, G. & Filmer, D. (1966). Comparison of experi-

mental binding data and theoretical models in proteins containing subunits. *Biochemistry, Easton*, **5**, 365–385.

84. Stadtman, E. R. (1966). Allosteric regulation of enzyme activity. *Adv. Enzymol.* **28**, 41–154.

85. Koshland, D. E. & Neet, K. E. (1968). The catalytic and regulatory properties of enzymes. *A. Rev. Biochem.* **37**, 359–410.

86. Kirschner, K. (1967). Temperature-jump relaxation kinetics with an allosteric enzyme: Glyceraldehyde-3-phosphate dehydrogenase in regulation of enzyme activity and allosteric interactions. In *Regulation of Enzyme Activity and Allosteric Interactions*. Proc. 4th Meeting of the Federation of European Biochemical Societies, Oslo. Ed. by Kvanne, E. & Pihl, A., pp. 39–58. London: Academic Press; Oslo: Universitetsforlaget.

87. Conway, A. & Koshland, Jr. D. E. (1968). Negative cooperativity in enzyme action. The binding of diphosphopyridine nucleotide to glyceraldehyde 3-phosphate dehydrogenase. *Biochemistry, Easton*, **7**, 4011–4022.

88. Kastenschmidt, L. L., Kastenschmidt, J. & Helmreich, E. (1968). The effect of temperature on the allosteric transitions of rabbit skeletal muscle phosphorylase *b*. *Biochemistry, Easton*, **7**, 4543–4556.

89. Helmreich, E. & Cori, C. F. (1964). The role of adenylic acid in the activation of phosphorylase. *Proc. natn. Acad. Sci. U.S.A.* **51**, 131–138.

90. Buc, H. (1967). On the allosteric interaction between 5′-AMP and orthophosphate on phosphorylase *b*. Quantitative kinetic predictions. *Biochem. biophys. Res. Commun.* **28**, 59–64.

91. Buc, M. H. & Buc, H. (1967). Allosteric interactions between AMP and orthophosphate sites on phosphorylase *b* from rabbit muscle. In *Regulation of Enzyme Activity and Allosteric Interactions*. Proc. 4th Meeting of the Federation of European Biochemical Societies, Oslo. Ed. by Kvanne, E. & Pihl, A., pp. 109–130. London: Academic Press; Oslo: Universitetsforlaget.

92. Madsen, N. B. & Shechosky, S. (1967). Allosteric properties of phosphorylase *b*. II. Comparison with a kinetic model. *J. biol. Chem.* **242**, 3301–3307.

93. Madsen, N. B. (1961). The inhibition of glycogen phosphorylase by uridine diphosphate glucose. *Biochem. biophys. Res. Commun.* **6**, 310–315.

94. Gerhart, J. C. & Pardee, A. B. (1962). The enzymology of control by feedback inhibition. *J. biol. Chem.* **237**, 891–896.

95. Graves, D. J., Scharfenberg-Mann, S. A., Philip, G. & Oliveira, R. J. (1968). A probe into the catalytic activity and subunit assembly of glycogen phosphorylase. *J. biol. Chem.* **243**, 6090–6098.

96. Wang, J. H. & Tu, J.-I. (1970). Allosteric properties of glutaraldehyde-modified glycogen phosphorylase *b*. *J. biol. Chem.* **245**, 176–182.

97. Wang, J. H. & Graves, D. J. (1964). The relationship of the dissociation to the catalytic activity of glycogen phosphorylase *a*. *Biochemistry, Easton*, **3**, 1437–1445.

98. Metzger, B., Helmreich, E. & Glaser, L. (1967). The mechanism of action of skeletal muscle phosphorylase *a* by glycogen. *Proc. natn. Acad. Sci. U.S.A.* **57**, 994–1001.

99. Whelan, W. J. & Cameron, M. P. (Eds.) (1964). *Control of Glycogen Metabolism*. Ciba Foundation Symposium. London: Churchill.

100. Whelan, W. J. (Ed.) (1968). *Control of Glycogen Metabolism*. Proc. 4th Meeting of the Federation of European Biochemical Societies, Oslo. London: Academic Press; Oslo: Universitetsforlaget.

101. Drummond, G. I. (1967). Muscle metabolism. *Fortschritte der Zool.* **18**, 360–429.
102. Helmreich, E. & Cori, C. F. (1965). Regulation of glycolysis in muscle. *Adv. Enz. Regul.*, **3**, 91–107.
103. Lowry, H. O., Passonneau, J. V., Hasselberger, F. X. & Schulz, D. W. (1964). Effect of ischemia on known substrates and cofactors on the glycolytic pathway in brain. *J. biol. Chem.*, **239**, 18–30.
104. Atkinson, D. E. (1965). Biological feedback control at the molecular level. *Science, N.Y.* **150**, 851–857.
105. Larner, J., Villar-Palasi, C., Goldberg, N. D., Bishop, J. S., Huijing, F., Wenger, J. I., Sacks, H. & Brown, N. B. (1968). Hormonal and nonhormonal control of glycogen synthesis—Control of transferase phosphatase and transferase I kinase. *Adv. Enz. Regul.* **6**, 409–423.
106. Passonneau, J. V. & Lowry, O. H. (1964). The role of phosphofructokinase in metabolic regulation. *Adv. Enz. Regul.* **2**, 265–274.
107. Leloir, L. F. (1964). Role of uridine diphosphate glucose in the synthesis of glycogen. In *Control of Glycogen Metabolism*, pp. 68–81. Ed. by Whelan, W. J. & Cameron, M. P. Ciba Foundation Symposium. London: J. A. Churchill, Ltd.
108. Mommaerts, W. F. H. M., Illingworth, B., Pearson, C. M., Guillory, R. J. & Seraydarian, K. (1959). A functional disorder of muscle associated with the absence of phosphorylase. *Proc. natn. Acad. Sci. U.S.A.* **45**, 791–797.
109. Schmid, R. & Mahler, R. (1959). Chronic progressive myopathy with myoglobinuria: Demonstration of a glycogenolytic defect in the muscle. *J. clin. Invest.* **38**, 2044–2058.
110. Lyon, J. B. & Porter, J. (1963). The relation of phosphorylase to glycogenolysis in skeletal muscle and heart of mice. *J. biol. Chem.* **238**, 1–11.
111. Krebs, E. G. & Fischer, E. H. (1955). Phosphorylase activity of skeletal muscle extracts. *J. biol. Chem.* **216**, 113–120.
112. Gerbach, E., Deuticke, B. & Dreisbach, R. H. (1963). Der Nucleotidabbau im Herzmuskel bei Sauerstoffmangel und seine mögliche Bedeutung fur die Coronardurchblutung. *Naturwissenschaften*, **50**, 228–229.
113. Morgan, H. E. & Parmeggiani, A. (1964). Regulation of glycogenolysis in Muscle III. Control of muscle glycogen phosphorylase. *J. biol. Chem.* **239**, 2440–2445.
114. Lyon, J. B., Porter, J. & Robertson, M. (1967). Phosphorylase *b* kinase inheritance in mice. *Science, N.Y.* **155**, 1550–1551.
115. Danforth, W. H. & Lyon, J. B. (1964). Glycogenolysis during tetanic contraction of isolated mouse muscle in the presence and absence of phosphorylase *a*. *J. biol. Chem.* **239**, 4047–4050.
116. Hurd, S. S., Teller, D. & Fischer, E. H. (1966). Probable formation of partially phosphorylated intermediates in the interconversions of phosphorylase *a* and *b*. *Biochem. biophys. Res. Commun.* **24**, 79–84.
117. Fischer, E. H., Hurd, S. S., Koh, P., Seery, V. L. & Teller, D. C. (1968). In *Control of Glycogen Metabolism*. Proc. 4th Meeting of the Federation of European Biochemical Societies, Oslo. Ed. by Whelan, W. J., pp. 19–33. London: Academic Press; Oslo: Universitetsforlaget.
118. Helmreich, E., Michaelides, M. C., Cori, C. F. (1967). Effect of substrates and a substrate analog on the binding of 5′-adenylic acid to muscle phosphorylase *a*. *Biochemistry, Easton*, **6**, 3695–3710.

119. Krebs, E. G., Graves, D. J. & Fischer, E. H. (1959). Factors affecting the activity of muscle phosphorylase *b* kinase. *J. biol. Chem.* **234**, 2867–2873.

120. Krebs, E. G., DeLange, R. J., Kemp, R. G. & Riley, W. D. (1966). Activation of skeletal muscle phosphorylase. *Pharmac. Rev.* **18**, 163–171.

121. Krebs, E. G., Love, D. S., Bratvold, G. E., Trayser, K. A., Meyer, W. L. & Fischer, E. H. (1964). Purification and properties of rabbit skeletal muscle phosphorylase *b* kinase. *Biochemistry, Easton*, **3**, 1022–1033.

122. Fischer, E. H. & Krebs, E. G. (1955). Conversion of phosphorylase *b* to phosphorylase *a* in muscle extracts. *J. biol. Chem.* **216**, 121–132.

123. Meyer, W. L., Fischer, E. H. & Krebs, E. G. (1964). Activation of skeletal muscle phosphorylase *b* kinase by Ca^{++}. *Biochemistry, Easton*, **3**, 1033–1039.

124. Huston, R. B. & Krebs, E. G. (1968). Activation of skeletal muscle phosphorylase kinase by Ca^{++} II. Identification of the kinase activating factor as a proteolytic enzyme. *Biochemistry, Easton*, **7**, 2116–2122.

125. Krebs, E. G., Huston, R. B. & Hunkeler, F. L. (1969). Properties of phosphorylase kinase and its control in skeletal muscle. *Adv. Enz. Regul.* **6**, 245–255.

126. Ozawa, E., Hosoi, K. & Ebashi, S. (1967). Reversible stimulation of muscle phosphorylase *b* kinase by low concentrations of calcium ions. *J. Biochem., Tokyo*, **61**, 531–533.

127. Ebashi, S., Endo, M. & Ohtsuki, I. (1969). Control of muscle contraction. *Q. Rev. Biophys.* **2**, 351–383.

128. DeLange, R. J., Kemp, R. G., Riley, W. D., Cooper, R. A. & Krebs, E. G. (1968). Activation of skeletal muscle phosphorylase kinase by adenosine triphosphate and adenosine 3′,5′-monophosphate. *J. biol. Chem.* **243**, 2200–2208.

129. Riley, W. D., DeLange, R. J., Bratvold, G. E. & Krebs, E. G. (1968). Reversal of phosphorylase kinase activation. *J. biol. Chem.* **243**, 2209–2215.

130. Walsh, D. A., Perkins, J. P. & Krebs, E. G. (1968). An adenosine 3′,5′-monophosphate-dependent protein kinase from rabbit skeletal muscle. *J. biol. Chem.* **243**, 3763–3765.

131. Soderling, T. R. & Hickenbottom, J. P. (1970). Inactivation of glycogen synthetase and activation of phosphorylase *b* kinase by the same cyclic 3′,5′-AMP dependent kinase. *Fedn Proc. Fedn Am. Socs exp. Biol.* **29**, 601 (Abstract).

132. Larner, J. & Sanger, F. (1965). The amino acid sequence of the phosphorylation site of muscle uridine diphosphoglucose α-1,4-glucan α-4-glucosyl transferase. *J. molec Biol.* **11**, 491–500.

133. Shapiro, B. M., Kingdon, H. S. & Stadtman, E. R. (1967). Regulation of glutamine synthetase, VII. Adenylyl glutamine synthetase: A new form of the enzyme with altered regulatory and kinetic properties. *Proc. natn. Acad. Sci. U.S.A.* **58**, 642–649.

134. Weiss, B. & Richardson, C. C. (1967). Enzymatic breakage and joining of deoxyribonucleic acid. *J. biol. Chem.* **242**, 4270–4278.

135. Little, J. W., Zimmerman, S. B., Oshinsky, C. K. & Gellert, M. (1967). Enzymatic joining of DNA strands II. An enzyme-adenylate intermediate in the DPN-dependent DNA ligase reaction. *Proc. natn. Acad. Sci. U.S.A.* **58**, 2004–2011.

136. Greengard, P., Rudolph, S. A. & Sturtevant, J. M. (1969). Enthalpy of hydrolysis of the 3′ bond of adenosine 3′,5′-monophosphate and guanosine 3′,5′-monophosphate. *J. biol. Chem.* **244**, 4798–4800.

137. Merlevede, W. & Riley, G. A. (1966). The activation and inactivation of phosphorylase phosphatase from bovine adrenal cortex. *J. biol. Chem.* **241**, 3517–3524.

138. Danforth, W. H., Helmreich, E. & Cori, C. F. (1962). The effect of contraction and of epinephrine on the phosphorylase activity of frog sartorius muscle. *Proc. natn. Acad. Sci. U.S.A.* **48**, 1191–1199.

139. Hurd, S. S. (1967). Phosphorylase phosphatase: Its purification and use in the study of the molecular properties of rabbit muscle glycogen phosphorylase. Ph.D. Thesis, University of Washington.

140. Pecararo, R. E. (1969). Phosphorylase phosphatase: Molecular properties and role in the control of glycogen metabolism. M.S. Thesis, University of Washington.

141. Meyer, F., Heilmeyer, L., Jr., Haschke, R. H., & Fischer, E. H., manuscript in preparation.

142. Heilmeyer, L., Jr., Meyer, F., Haschke, R. H., & Fischer, E. H., manuscript in preparation.

143. Haschke, R. H., Meyer, F., Heilmeyer, L., Jr., & Fischer, E. H., manuscript in preparation.

144. Yunis, A. A., Fischer, E. H. & Krebs, E. G. (1962). Comparative studies on glycogen phosphorylase IV. Purification and properties of rabbit heart phosphorylase. *J. biol. Chem.* **237**, 2809–2815.

145. Davis, C. H., Schliselfeld, L. H., Wolf, D. P., Leavitt, C. A. & Krebs, E. G. (1967). Interrelationships among glycogen phosphorylase isozymes. *J. biol. Chem.* **242**, 4824–4833.

146. Appleman, M. M., Krebs, E. G. & Fischer, E. H. (1966). Purification and properties of inactive liver phosphorylase. *Biochemistry, Easton*, **5**, 2101–2107.

147. Yunis, A. A. & Krebs, E. G. (1962). Comparative studies on glycogen phosphorylase II. Immunological studies on rabbit and human skeletal muscle phosphorylase. *J. biol. Chem.* **237**, 34–39.

148. Karpatkin, S. & Langer, R. M. (1969). Human platelet phosphorylase. *Biochim. biophys. Acta*, **185**, 350–359.

149. Metzger, B. E., Glaser, L., and Helmreich, E. (1968). Purification and properties of frog skeletal muscle phosphorylase. *Biochemistry, Easton*, **7**, 2021–2036.

150. Cohen, P., unpublished results.

151. Cowgill, R. W. (1959). Lobster muscle phosphorylase: Purification and properties. *J. biol. Chem.* **234**, 3146–3153.

152. Assaf, S. A. & Graves, D. J. (1969). Structural and catalytic properties of lobster muscle glycogen phosphorylase. *J. biol. Chem.* **244**, 5544–5555.

153. Fosset, M. and Nielsen, L. D., unpublished results.

154. Shepherd, D. & Segel, I. H. (1969). Glycogen phosphorylase of *Neurospora crassa*. *Archs Biochem. Biophys.* **131**, 609–620.

155. Chen, G. S. and Segel, I. H. (1968). Purification and properties of glycogen phosphorylase from *Escherichia coli*. *Archs Biochem. Biophys.* **127**, 175–186.

156. Schwartz, M. & Hofnung, M. (1967). La maltodextrine phosphorylase d'*Escherichia coli*. *Eur. J. Biochem.* **2**, 132–145.

157. Lee, Y. P. (1960). Potato phosphorylase I. Purification, physicochemical properties and catalytic activity. *Biochim. biophys. Acta*, **43**, 18–24.

158. Fukui, T. & Kamogawa, A. (1969). Structure and function of potato α-glucan

68   E. H. FISCHER, A. POCKER AND J. C. SAARI

phosphorylase; comparison with the muscle enzyme. *J. Jap. Soc. Starch Sci.* **17**, 117–129.

159. Sutherland, E. W., Wosilait, W. D. (1956). The relationship of epinephrine and glucagon to liver phosphorylase I. Liver phosphorylase; preparation and properties. *J. biol. Chem.* **218**, 459–468.

160. Harris, C. E. (1969). I. The molecular characteristics of yeast aldolase; II. The estimation of sequence homology from amino acid composition of evolutionarily-related proteins. M.Sc. Thesis, University of Washington.

161. Henion, W. F. & Sutherland, E. W. (1957). Immunological differences of phosphorylases. *J. biol. Chem.* **224**, 477–488.

162. Chester, V. E. (1964). Comparative studies on the dissimilation of reserve carbohydrate in four strains of *Saccharomyces cerevisiae. Biochem. J.* **92**, 318–323.

163. Blake, C. C. F., Johnson, L. N., Mair, G. A., North, A. C. T., Phillips, D. C. & Sarma, V. R. (1967). Crystallographic studies of the activity of hen egg-white lysozyme. *Proc. R. Soc. B*, **167**, 378–388.

164. Butcher, R. W., Robison, G. A., Hardman, J. G. & Sutherland, E. W. (1968). The role of cyclic AMP in hormone actions. *Adv. Enz. Regul.* **6**, 357–389.

165. Robison, G. A., Butcher, R. W. & Sutherland, E. W. (1968). Cyclic AMP. *A. Rev. Biochem.* **37**, 149–174.

166. Huxley, H. E. (1969). The mechanism of muscular contraction. *Science, N.Y.* **164**, 1356–1366.

167. Peachey, L. D. (1968). Muscle. *A. Rev. Physiol.* **30**, 401–440.

Proinsulin, a Biosynthetic Precursor of Insulin

P. T. GRANT and T. L. COOMBS

Natural Environment Research Council, Fisheries
Biochemical Research Unit, University of Aberdeen,
Aberdeen, Scotland

I. Introduction

The cleavage of biologically active polypeptides and proteins from preformed and inactive precursors, by a restricted or limited hydrolysis involving only a proportion of the potentially susceptible peptide bonds in the precursor, is an extracellular process that effects the activation of enzymes and complement, coagulates blood and milk[1] and releases small peptides with intense pharmacological activity.[2] Restricted intracellular proteolysis may also be involved in the control of metabolic processes and recent work initiated in 1967 by D. F. Steiner of the University of Chicago indicates that insulin in the β cells of the Islets of Langerhans is specifically derived from a larger polypeptide precursor, termed proinsulin. Although the conversion of a precursor to an active hormone may be an intracellular event peculiar to the β cell, this process may also occur in other cells, such as the α cells of the endocrine pancreas and the cells of the pituitary gland, that specialize in the formation, storage and export of other small polypeptide hormones.[3] This essay is primarily concerned with the nature and properties of proinsulin in relation to the overall sequence of events in the secretory cycle of the β cell. This sequence of events and the control mechanisms exercised by both intracellular and extracellular signals have an obvious bearing on our understanding of the state of diabetes mellitus, which in many cases can be attributed to an absolute or relative inadequacy of β cell function.[4]

II. Biosynthesis of Insulin

A. β CELLS AND THE ISLETS OF LANGERHANS

The Islets of Langerhans are discrete islands of endocrine tissue embedded in a predominant mass of exocrine cells comprising the pancreas of most animals. The islet tissue is mainly composed of α cells that secrete glucagon and β cells that secrete insulin, as well as other cell types of undefined function that together constitute about 1% of the total pancreatic mass in mammals. For experimental studies it is essential to separate the islets from the mass of exocrine cells and their potent digestive secretions. Microdissection has been used, but a simpler and less tedious method has been developed by Lacy and Kostianovsky.[5] In this procedure, the pancreatic tissue is disrupted by injection of saline into the main pancreatic duct and the chopped gelatinous mass digested with collagenase (EC 3.4.4.19) to separate endocrine cells from fragments of connective tissue. The discrete islets, containing about 50–60% of β cells, are unaffected by this treatment and can be recovered by differential centrifugation since they sediment more readily than the exocrine elements.

More restricted sources, such as human tumours of the β cell or the pancreatic tissue of obese hyperglycemic mice, have also been used, and preparations containing 80–90% of β cells were obtained by simple dissection. The islet tissue of certain fishes is an anatomically discrete mass of mainly β cells, and as we shall see the isolated tissue has been extensively used as a convenient model system to study insulin biosynthesis.

B. STRUCTURE OF INSULINS AND A PROBLEM IN BIOSYNTHESIS

Insulin is stored in membrane-limited granules within β cells, as will be discussed later (Fig. 5), and this accounts for the high concentration of preformed insulin that amounts to about 10–15% of the dry weight of cells.[6, 7] All insulins are composed of two polypeptide chains; an A chain of 21 amino acid residues containing an intra-chain disulphide bond, connected by two inter-chain disulphide bridges to a B chain containing either 29 or 30 amino acid residues. There is a considerable degree of species variation in the primary structure of the chains and this is particularly marked in the insulins of the guinea-pig and certain fishes.[8] A comparison of the complete or partial sequences of 20 insulins indicates that about half the residues are invariant (Fig. 1). The nature and location of these invariant residues is in general agreement with the essential features of bovine insulin determined by the effects of chemical modification on biological activity.[9] The intra-chain and inter-chain disulphide bonds, therefore, have an essential but not exclusive role in the activity of these insulins.

A feature of insulin is the formation at some point in the biosynthetic sequence of both the intra-chain and inter-chain disulphide bonds by the correct pairing of the invariant cysteine residues in the polypeptide chains. Since insulin biosynthesis is inhibited by both puromycin[11, 12] and cycloheximide,[13] a conventional RNA-directed mechanism is involved. However, two quite distinct biosynthetic pathways may be envisaged.

At first sight the simplest pathway would be the separate and parallel biosynthesis of each polypeptide chain with a subsequent directed mechanism for disulphide bond formation from the correct pairs of cysteine residues (Fig. 2). A directed mechanism must be involved since, in theory, the random formation of disulphide bonds would result in polymers as well as twelve isomeric structures each containing two polypeptide chains and only one of these isomers is the active hormone. This pathway would be similar to that involved in the biosynthesis of immunoglobulins[14] and the directed mechanism could involve a template, an enzyme, or even the unique complementary conformation of each polypeptide chain one to the other. However, the yields of insulin obtained experimentally by either spontaneous[15] or enzymic[16] recombination of the reduced chains are low except under special conditions that are unlikely to exist in the intact β cell. The principal products were usually insoluble polymers containing both A and B chains.

The second possible biosynthetic pathway was suggested by Anfinsen and his co-workers,[17] from a comparison of the behaviour of insulin with that of small proteins consisting of a single polypeptide chain with intra-chain disulphide bonds, such as ribonuclease (EC 2.7.16) and chymotrypsinogen. In mildly alkaline buffers containing dissolved

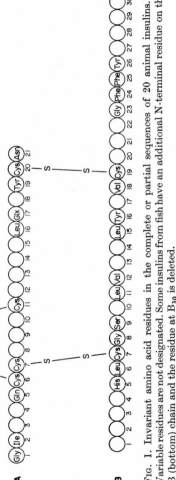

Fig. 1. Invariant amino acid residues in the complete or partial sequences of 20 animal insulins.[10] Variable residues are not designated. Some insulins from fish have an additional N-terminal residue on the B (bottom) chain and the residue at B_{30} is deleted.

oxygen the chemically reduced and unfolded forms of these proteins tended to reassume spontaneously the native conformation and reform

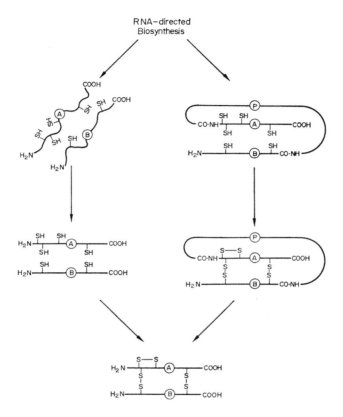

FIG. 2. A hypothetical scheme to illustrate two possible mechanisms of insulin formation from either the separate A and B chains, or a precursor polypeptide. —(A)— —(B)— represent the amino acid sequences of the A and B chains of insulin and —(P)—, represents a peptide segment of the precursor that connects the carboxy-terminus of the B chain to the amino-terminus of the A chain. An alternative structure of the precursor (not shown) would be the connection by the peptide segment, —(P)— of the carboxy-terminus of the A chain to the amino-terminus of the B chain. For details see text.

the correct disulphide bonds as judged by various physical and chemical criteria. Similarly, the configuration of these native proteins was not significantly affected by the presence of an enzyme that catalysed disulphide–sulphydryl exchange reactions. In contrast, after the selective cleavage of a peptide bond at a position in the chain of ribonuclease internal to a disulphide bond, or after the activation of chymotrypsinogen to α-chymotrypsin (three separate polypeptide chains containing inter-chain disulphide bonds), these multi-chain forms behaved like insulin and randomly polymerized or separated into individual chains in the presence of the enzyme that catalysed disulphide–sulphydryl exchange reactions. These results indicated that the conformation of both ribonuclease and chymotrypsinogen is dictated solely by the non-covalent interactions of the amino acid side chains permitted by the primary structure. Thus the native protein is envisaged as being in a state of dynamic equilibrium with other conformational variants representing higher thermodynamic energy states whose equilibrium concentrations are probably below the limits of detection. In these proteins the disulphide bonds are additional and essential stabilizing forces. In contrast, it may be inferred that the disulphide bonds are the primary stabilizing force in the native form of the insulin molecule. On the basis of these results Anfinsen and his co-workers[17] suggested that the A and B chains of insulin might be biosynthesized as constituent parts of a larger single polypeptide chain (see Fig. 2) whose native conformation would dictate the subsequent formation of disulphide bonds between the correct cysteine residues. They referred to this precursor as proinsulin and concluded that insulin would be derived from this biosynthetic precursor by the selective hydrolysis of peptide bonds.

C. THE IDENTIFICATION OF PROINSULIN AS THE PRIMARY PRODUCT OF BIOSYNTHESIS

One of the first experimental studies on the mechanism of insulin biosynthesis was made by Humbel,[11] who attempted to distinguish between the separate and parallel biosynthesis of the A and B chains and the biosynthesis of a single polypeptide chain precursor, on the assumption that in either event the synthesis would be initiated and proceed from the N-terminal end of the peptide chain or chains, as had been previously shown by Dintzis and co-workers[19] for the direction of haemoglobin biosynthesis. Slices of anglerfish islets were incubated with [^{3}H]proline for 15 min and the specific activities of proline residues located at A_9, B_2 and B_{28} of insulin were determined. The observed labelling pattern, where $B_{28} > A_9 > B_2$ (see Fig. 1), was inconsistent with the derivation of insulin from a single chain precursor and provided indirect evidence for the concept of a separate and parallel synthesis of the A and B chains. However, as we shall see, there is now direct evidence for the existence and role of proinsulin in the β cells of several species including the anglerfish[18] and it is difficult to envisage that both pathways of biosynthesis (see Fig. 2) are present in the same cell. It remains possible that Humbel's interpretation of his results was complicated by the presence in anglerfish islets of more than one prolyl transfer-RNA, each containing anticodons specific for certain of the four proline codons and each utilizing different metabolic pools of proline with different turnover rates. For example, two different species of arginyl transfer-RNA transfer arginine into different positions of rabbit haemoglobin in a way that suggests that the three arginine residues of the α chain correspond to three different codons.[20]

Although insulins have a marked tendency to aggregate into soluble polymers in neutral solution, they exist as monomers (mol.wt. about 6000) in dilute acid, and gel-filtration on Sephadex G-50 in M-acetic acid has been extensively used to separate biosynthetic products involved in insulin formation. This procedure was used by Steiner and Oyer[21] to demonstrate that slices of a human islet cell adenoma incorporated either [^{3}H]leucine or [^{3}H]phenylalanine into a single polypeptide (mol.wt. about 10,000). They concluded that it was an insulin precursor, since (1) it reacted strongly with antiserum to insulin, (2) it had a higher specific radioactivity than the insulin obtained from the tissue slice, and (3) it was progressively converted by trypsin (EC 3.4.4.4) to a product that had properties similar to that of insulin on gel-filtration. Further work has confirmed the occurrence of an insulin precursor, or proinsulin, in the islet tissue of the rat,[13,21] foetal calf,[22] hamster,[23] cod[12] and anglerfish.[18]

Proinsulin is present in β cells in low concentrations (about 5% of that of insulin) and a precursor-product relationship has been demonstrated.[12,24] Grant and Reid[12] showed that cod proinsulin had a negligible biological activity in the epididymal fat-pad bioassay, but on treatment with trypsin there was a progressive conversion to an active insulin-like material. In cod, but not apparently in rat islets, the conversion of proinsulin to insulin is inhibited by O-ethyl-O-(p-nitrophenyl)-phenylpropylphosphonate and di-isopropylfluorophosphate.[12] These compounds are known to be stoicheiometric inhibitors of enzymes that contain a unique serine as part of the catalytic centre, and a trypsin-like enzyme may be involved in the conversion process within the β cell.

Proinsulin tends to co-crystallize with insulin and the discovery by Chance, Ellis and Bromer[25] that small amounts of proinsulin are present in crystalline commercial preparations has made possible the isolation of sufficient ox and pig proinsulins for a more detailed study of their properties.

III. Nature and Properties of Proinsulin

A. STRUCTURE AND STABILITY

Both ox and pig proinsulin are similar single-chain polypeptides (see Fig. 3) where the inclusive amino acid sequences of the A and B chains of insulin are joined together by a connecting peptide segment (connecting peptide) that links the carboxy-terminal alanine of the B chain to

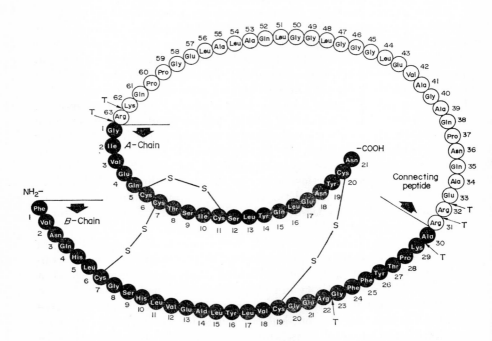

Fig. 3. Provisional primary amino acid sequence of pig proinsulin. The A and B chains of insulin are depicted by dark circles. ← T, indicates peptide bonds that fulfil the specificity requirements of trypsin. Adapted from Chance, Ellis and Bromer.[25]

the amino-terminal glycine of the A chain.[25, 26] Although the basic amino acid residues at each end of the connecting peptide are the same in the two species, the connecting peptides of pig and cod[12] proinsulins contain three amino acid residues more than ox proinsulin and the amino acid compositions show a considerable degree of variation (Table 1).

In contrast to ox and rat insulins, both the corresponding proinsulins, after reduction of all disulphide bonds can be reoxidized under mildly alkaline conditions to yield a product (about 70% of a theoretical maximum) that by a number of criteria is identical with the native

TABLE 1

Comparison of amino acid composition of the connecting peptides from ox,[26] pig[25] and cod[12, 27] proinsulins

	Ox	Pig	Cod
Ala	3	5	4
Arg	3	3	2
Asp	0	1	1
Gln	2	4	⎫
Glu	4	3	⎬ 9
Gly	8	7	1
Leu	3	5	3
Lys	1	1	4
Met	0	0	1
Pro	4	3	3
Ser	0	0	3
Thr	0	0	2
Val	2	1	0
	30	33	33

polypeptide.[26, 28] These crucial experiments suggest that the conformation of the nascent proinsulin after its formation is dictated by its primary amino acid sequence and that its native conformation ensures the pairing of cysteine residues for the subsequent formation of the correct disulphide bonds as ancillary stabilizing forces. The directing role of the connecting peptide in the stabilization of the native conformation of reduced proinsulin is obscure, but may be related to the invariant residues present in both the connecting peptides of pig and ox proinsulin. The directing role of the connecting peptide is of more than a theoretical interest, since it has been pointed out by Steiner[26] that "a random mutation occurring in the connecting peptide might alter the folding efficiency of proinsulin and is relevant to the consideration of possible genetic defects in proinsulin which might be responsible for heritable forms of diabetes".

B. IMMUNOLOGICAL PROPERTIES

The major antigenic determinants in insulins are thought to be located in the regions occupied by amino acid residues at positions 10–21 inclusive of the A chain and positions 1–8 inclusive of the B chain[29] (see Fig. 1). An additional antigenic determinant is also located in the connecting peptide of proinsulin.[30–32] The occurrence of this locus, together with the two other loci that are common to both insulin and proinsulin, is a matter of some consequence to the specificity of radio-immunoassay procedures used to determine insulin, or more precisely, "immunoreactive insulin" in tissue and body fluids.

Ox, pig and human insulins cross react with an antibody prepared against ox insulin but the corresponding proinsulins showed some degree of species specificity towards an antiserum prepared against ox pro-insulin.[30] This is in agreement with the differences in structure between the connecting peptides of ox and pig proinsulin and indicates that the locus on the connecting peptide is a major antigenic determinant in the proinsulin molecule. At the same time, it may also be concluded that the unknown structure of the connecting peptide of human proinsulin must differ in some respects from that of ox proinsulin.

C. BIOLOGICAL ACTIVITY BEFORE AND AFTER DIGESTION WITH TRYPSIN

Proinsulins from various species have only a low insulin-like activity on isolated tissue preparations and in bioassays using whole animals.[12, 25, 33, 34] It is still, however, an open question whether this is due to an intrinsic activity of proinsulin itself rather than to a partial conversion to insulin by tissue proteases.

The major product of tryptic digestion of pig[25] or ox proinsulin[26, 35] at an enzyme/substrate ratio of less than 1:100 is mainly dealanine insulin (insulin minus the carboxy-terminal alanine of the B chain) which has the same biological activity as insulin. A detailed study of the intermediate products of digestion of pig proinsulin by Chance[34] indicated that the six potentially sensitive bonds (see Fig. 3) were not split at the same rate. The bonds at $-\text{Arg}_{63}-\text{Gly}_{1}-$ and $-\text{Arg}_{32}-\text{Glu}_{33}-$ are rapidly hydrolysed and the basic intermediate insulin-like product is more slowly hydrolysed at $-\text{Lys}_{29}-\text{Ala}_{30}-$ to yield dealanine insulin. The

bond at $-$Arg$-$Gly$-$ in insulin is known to be only slowly hydrolysed[34]
(with subscripts 22 and 23 under Arg and Gly)
as are peptide bonds formed by two basic residues[8, 36] or that formed
by an amino-terminal basic residue.[37] On the basis of these model
studies it has been suggested that the conversion of mammalian proinsulin to insulin in the β cell might be effected by the combined action
of two enzymes with specificities that correspond to trypsin and carboxypeptidase B (EC 3.4.2.2) respectively.

IV. Secretory Cycle of the β Cell

A. ULTRASTRUCTURE

Current views concerning the secretory pathway and functional
interrelationships of cell organelles in the β cell are that there is a qualitative similarity to that established by Palade, Siekevitz and co-workers
in their now classical studies on the secretion of zymogens and enzymes
from the cells of the exocrine pancreas.[38, 39]

The postulated pathway involves first a transfer of the newly synthesized protein from the ribosomes into the cisternae formed by the
membranes of the endoplasmic reticulum. The Golgi complex of the β
cell is usually well developed, and some investigators believe it to be
involved in the packaging of protein into the characteristic electron-
lucent vesicles.[40] Others suggest that these vesicles arise directly from
elements of the endoplasmic reticulum that are pinched off and lose
their attached ribosomes (Fig. 4). Whatever the origin of the electron-
lucent vesicles, they are usually considered to be immature granules
that have not yet concentrated their contents to the point of forming a
crystalline matrix to form a mature electron-dense secretory granule[41]
(Fig. 5).

The fine structure of mature β granules varies with the species, but the
electron-dense matrix is thought to consist mainly of crystals of zinc–
insulin polymers. The migration of β granules through the cell is an
energy-dependent process and the attractive suggestion has been made
by Lacy, Howell, Young and Fink[42] that an internal cytoskeleton
composed of microtubules and contractile actomyosin-like filaments
may link secretory granules to the cytoplasmic membrane to provide
a pathway of controlled translocation within the β cell. In support of
this idea it was shown that colchicine inhibited the secretion of insulin
from rat islets. Colchicine is well known to bind strongly to the microtubules of both plant and animal cells and these microtubules have been
implicated in the motile systems of cilia and flagella, in the maintenance
of cell asymmetries and with the movement of cytoplasmic particles

within cells.[43] The movement of vesicles by microtubules may have some analogy with the sliding-filament theory of muscular contraction.[44]

The process of discharge of granules from the cell may involve the coalescence of the membrane-limited granules with the cell membrane where the crystalline product is extruded and undergoes extracellular dissolution. Essentially the same mechanism of release appears to be

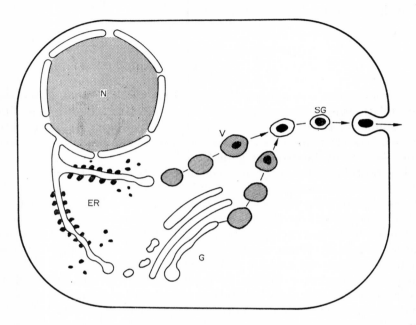

FIG. 4. Diagram illustrating current concepts of the intracellular sequence of events in the β cell culminating in the extracellular release of an insulin granule. Nucleus, N; endoplasmic reticulum, ER; Golgi apparatus, G; vesicle, V; and secretory membrane-limited granule, SG. For description see text.

common to a wide variety of endocrine and neuroendocrine cell types that form secretory droplets or granules that are visible in the electron microscope.[39]

B. THE CONVERSION OF PROINSULIN TO INSULIN

Two lines of evidence support the idea that the conversion of pro-insulin to insulin is an intracellular process in the β cell. These are: (1) that the relative amount of proinsulin in the β cells of several species is about 5% of that of insulin[26] and there is a rapid synthesis and translocation of proinsulin from the endoplasmic reticulum into membrane-limited vesicles or granules with a concomitant conversion to insulin;[45,46] and (2) that intact islet cells release roughly equimolar amounts of insulin and free connecting peptide into the suspending medium,[31] and do not convert ^{131}I-labelled proinsulin to insulin when it is added to the medium.[26] The central problem concerning the nature and intracellular location of the converting enzyme or enzymes remains.

Much of our knowledge about the mechanisms of intracellular digestion is due to de Duve and his co-workers[47] and has been concerned with the acid hydrolases present in the lysosomes of many types of cells. The action of the acid cathepsins of lysosomes on proteins and peptides[48] is a relatively non-specific process of digestion and the only evidence for the role of lysosomes in the β cell[73] and other secretory cells is concerned with the disposal of secretory granules produced in excess of demand. This process for the digestive disposal by a cell of its excess secretory products has been termed "crinophagy" by de Duve.[47]

Clearly, at least one of the enzymes that catalyses the limited or restricted cleavage of proinsulin to form insulin in the β cell must possess a high degree of specificity that is similar, if not identical, to that of trypsin. Intracellular proteases of this specificity are almost unknown, but enzymes of this type with an alkaline pH optimum have been reported to be present in human leukocytes[74] and in the secretory granule fraction of rat pituitary glands.[75] However, a study of the nature of proteolytic enzymes in the subcellular fractions derived from β cell preparations of the cod[49] would indicate that essentially all the activity in the freshly isolated microsome and vesicle-granule fractions was present as a membrane-bound precursor whose properties resembled those of trypsinogen derived from exocrine cells. A detailed characterization of this membrane-bound precursor protein is clearly required. Even if it proves to be indistinguishable from trypsinogen, the possibility remains that it could still be a constitutive protein of the β cell or other endocrine cells, since there is some evidence to suggest that both endocrine and exocrine cells evolve from a common ancestral cell in the early stages of embryonic development of the mammalian pancreas.[50] Moreover, the apparently normal pancreas of mature animals has been reported[51] to contain "mixed" or "intermediate" cell types that could represent a transition stage in the interconversion of exocrine and endocrine cells.

The site of conversion of proinsulin to insulin in the cod β cell appears to be restricted to a subcellular fraction containing mainly vesicles and granules. The conversion that occurs during the *in vitro* incubation of this isolated fraction was not affected by the presence of a polymeric trypsin inhibitor (mol.wt. 28,000) in the suspending medium, but was almost completely abolished when the particulate elements were lysed with deoxycholate or treated with DFP before incubation.[49] These results would suggest that both proinsulin and the DFP-sensitive enzyme that catalyses the conversion are present and are contained in membrane-limited vesicles or granules. There is no direct evidence that the membrane-bound trypsinogen-like protein also present in this subcellular fraction is the precursor of the converting enzyme, but it is tempting to speculate that the β cell may, with advantage, synthesize both an active hormone and a potent proteolytic enzyme in the form of their inactive precursors. This would enable both precursors to be packaged and contained in membrane-limited vesicles before an activation mechanism results in the eventual formation of the active hormone.

C. A FUNCTIONAL ROLE FOR ZINC

Islet cells of many species are rich in zinc. Histochemical evidence would indicate that the metal is mainly present in vesicles and secretory granules.[52] Although there is no conclusive proof, it seems probable that the electron-dense material in the granule consists of zinc–insulin crystals. High-resolution electron microscopy of a negatively stained secretory granule (Fig. 5) shows that it consists of a highly ordered structure with a periodicity of about 50 Å. This is of the same order of magnitude as that obtained from X-ray diffraction measurements of the zinc–insulin hexamer present in insulin crystals.[53] Zinc is essential for the crystallization of insulin, two atoms of zinc being firmly co-ordinated per insulin hexamer and the hexameric units condense to form the insulin crystal. X-ray diffraction studies have shown that each zinc atom in the hexamer is co-ordinated to the imidazolyl groups of three histidine residues at position B_{10} of each insulin molecule and to three other atoms, probably the oxygens of co-ordinated water molecules.[53] In contrast, the proinsulins of ox, pig and cod do not appear to form insoluble zinc salts or crystals[54] and there must be a quantitative difference in the nature or degree of zinc binding between proinsulin and the corresponding insulin.

Insulin containing small quantities of zinc is very resistant to digestion by trypsin,[9,55] carboxypeptidase[56] (EC 3.4.2.1) and leucine amino-peptidase[57] (EC 3.4.1.1), and although it has not been tested experimentally, the possibility must be considered that zinc has a functional role in the enzymic conversion of proinsulin to insulin within the β cell. It is probable that nascent insulin would chelate strongly with zinc, so that degradation is minimized and the complete conversion of

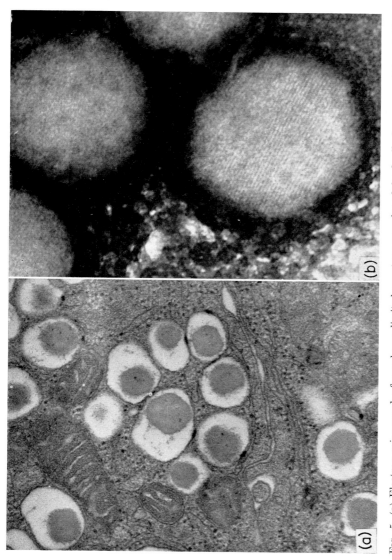

Fig. 5. (a) Electron micrograph of membrane-limited secretory granules of the β cell of pancreas of the rat. The angular profile of the granule and the large space between the granule and its limiting membrane are characteristic patterns of these granules in the rat. ×48,000. (b) Electron micrograph of a granule from the rat β cell, negatively stained with phosphotungstic acid. ×200,000. Periodicity, 50 Å. Both micrographs from Greider, Howell and Lacy.[41]

[*Facing p. 82*].

proinsulin to insulin would be favoured in a thermodynamic sense by the insoluble nature of the zinc–insulin polymers:

$$\text{Proinsulin} \rightarrow \begin{array}{c} \text{insulin} \\ + \text{ connecting peptide} \end{array} \xrightarrow{\text{Zn}^{2+}} \begin{bmatrix} \text{zinc–insulin} \\ \text{hexamer} \end{bmatrix} \xrightarrow{\text{Zn}^{2+}} \begin{array}{c} \text{zinc–insulin} \\ \text{insoluble polymer} \end{array}$$

The stabilization of a protein by ligands against the action of certain proteolytic enzymes has been observed for staphylococcal nuclease[58]

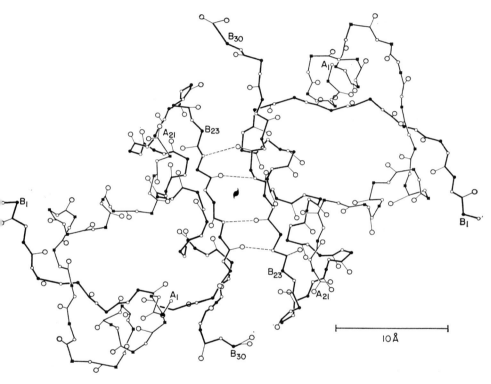

Fig. 6. A drawing that illustrates the spatial relationships of the peptide backbones of the A and B chains in the dimer of pig insulin. Drawn from X-ray crystallographic data.[76] ●, ■, α carbon atoms in the polypeptide backbone of the B and A chains respectively. ○, nitrogen, and ⊙, oxygen atoms, in polypeptide backbones. ----, hydrogen bonds.

(EC 3.1.4.7) as well as for insulin. The digestion of native proteins is usually presumed to proceed through the dynamic equilibrium of the native form with other conformational variants. A ligand that binds only to the native form would stabilize it with respect to other variants, including the denatured state so that an inhibition of the digestion by

proteolytic enzymes would be expected. In addition, the three-dimensional structure of the zinc-insulin hexamer is such that the trypsin-sensitive bond, $-Arg-Gly-$, present in the B chain of insulin is partially
            22   23
buried in the interior part of the hexameric unit.[76] It seems improbable that trypsin could gain access to form an enzyme–substrate complex. This may be illustrated by reference to the close contact found in models, constructed from X-ray crystallographic data, between a large part of each B chain in the insulin dimer (Fig. 6) that forms part of the stable hexameric unit. These regions of each B chain are arranged in an antiparallel manner and appear to form a hydrogen-bonded close-fitting pattern (pleated sheet).[53,76]

D. CONTROL MECHANISMS

To maintain a state of homeostasis (stability of internal environment), it is perhaps self-evident that the rate of secretion of insulin from the β cell must be the subject of very precise control by the concentration of specific substances in the circulation and by the nervous system. Radioimmunoassay procedures developed by Yalow and Berson, and Hales and Randle have been a sensitive tool for the identification of a large number of factors that either stimulate or inhibit the release of insulin under *in vitro* conditions (Table 2). These factors may be presumed to act directly on the β cell, but the manner in which they individually or collectively effect the control of insulin secretion is not known. The interested reader is referred to recent comprehensive reviews[59–61] for a more detailed discussion.

A major factor controlling the release of insulin in most animals is the concentration of glucose in the suspending medium or blood. The response of the β cell to glucose can be modified by the presence of other hormones, drugs, and metal ions such as Ca^{2+}. The nutritional status can also affect the response of the β cell in the normal animal and it is also modified in obesity, pregnancy and diabetes. In the normal animal, when the blood-sugar level rises above the basal range of concentration, insulin secretion is stimulated, as determined by an increase in the "immunoreactive-insulin" level of the blood. The increased level of insulin in the circulation enhances glucose uptake by muscle and adipose tissue and inhibits hepatic gluconeogenesis so that the blood-glucose level falls and insulin secretion is diminished. This is a feedback mechanism and ensures that there is a reciprocal relationship in the normal animal between the rate of insulin secretion from the β cell and the level of circulating insulin in the blood. In addition, at a given concentration of blood glucose there is some evidence to suggest that changes in the level of insulin in the blood can effect a compensatory change in the rate of secretion of insulin from the β cell.[60] The response of the β cell, however, depends on the duration of the stimulus. If the β

TABLE 2

Effect of various substances on the secretion of insulin by isolated islet tissue or slices of pancreas[a]

Activators	
Sugars and related compounds	Glucose (M, D, R), mannose (R, Rb, H), xylitol (Rb, H), ribitol (R), ribose (R)
Amino acids	Arginine (M, R), leucine (M, R), phenylalanine (M, R), lysine (M, R)
Fatty acids	Butyrate (R, S, C), octanoate (R), acetoacetate (R, Rb, D), β-hydroxybutyrate (R, Rb, D)
Hormones	Secretin (D, Rb), glucagon (M, R, Rb, D, Dk)
Metal ions	Ca^{2+} (R), K^+ (R)
Drugs	Tolbutamide (M, R, Rb, D), theophylline (R), caffeine (R)
Nucleotide	Cyclic AMP (R)
Inhibitors	
Sugars and related compounds	Mannoheptulose (R, Rb), 2-deoxyglucose (R)
Uncoupling agents	2,4-Dinitrophenol (R, Rb)
Hormones	Insulin (R), growth hormone (M, R, D)
Drug	Diazoxide (R)

[a] Capital letters in parentheses indicate the source of tissue: man (M); rat (R); rabbit (Rb); dog (D); hamster (H); sheep (S); cow (C); duck (Dk).

cell is stimulated by glucose over a prolonged period there is a compensatory increase in the number of new β cells. For example, the continuous infusion of glucose to a rat for seven days can double the mass of islet tissue.[63]

Inhibitors of glucose metabolism in β cells and other tissues have been used in attempts to identify the metabolic pathway or metabolic product that acts as a trigger to regulate the secretion of insulin. Randle and others have found that glucose had no effect on secretion when the metabolism is inhibited by mannoheptulose or glucosamine (inhibitors of glucokinase, EC 2.7.1.1, and hexokinase, EC 2.7.1.2), by uncoupling agents such as 2,4-dinitrophenol and by anoxia.[59,62] Similarly, the stimulating effect of glucose on the biosynthesis of proinsulin and its conversion to insulin was inhibited by mannoheptulose.[64] These findings indicate that the first step (biosynthesis) and last step (transport and secretion process) of the overall sequence of events in the β cell are stimulated above the basal level by the metabolism of exogenous glucose. Both steps are energy-dependent processes and their relative rates may be directly dependent on the energy level of the β cell, i.e. the intracellular concentration of ATP or other high-energy compounds. The energy level of the β cell, in turn, would depend on the concentrations of metabolizable substrates (Table 2) that are available and permeable to the cell for oxidation by the tricarboxylic acid cycle and the pentose phosphate pathway. These observations leave open the question as to whether the individual events such as biosynthesis, packaging and vesicle formation, intracellular transport and secretion of granules are related in a manner analogous to a steady-state system so that stimulation of one stage results in a stimulation of the overall process. This does not, however, appear to be the case, since ribose and other substances stimulate the secretion of insulin in rat islets without apparently affecting the biosynthesis of proinsulin.[64] It is therefore probable that biosynthesis on one hand and intracellular transport and insulin release on the other are separate events in the β cell and subject to different regulatory mechanisms. A similar situation appears to be present in pancreatic exocrine cells, where the transport of zymogen granules is essentially unaffected in cells when protein synthesis is inhibited by cycloheximide.[67]

V. Proinsulin and "Big" Insulin in Serum

The theme that has been developed so far in this essay is that proinsulin is a transient biosynthetic intermediate in the formation of insulin by the β cell. It seems likely, however, that proinsulin is also secreted by the β cell under certain conditions. The amount secreted, relative to that of insulin, has not been determined quantitatively, so that the significance remains to be evaluated.

Proinsulin was detected by gel-filtration of serum or urine, to separate proinsulin and insulin, followed by the radioimmunoassay of each fraction obtained from the column of gel. The two immunoreactive components in serum corresponded in molecular size to proinsulin and insulin but were not further characterized, and have consequently been referred to as "big" insulin and "little" insulin, respectively.[65] The identity of "big" insulin with proinsulin has been further indicated by the use of antisera that reacted strongly with proinsulin and had only a weak[66] or negligible[32] cross-reaction with insulin. On the limited evidence available it is probable that significantly high levels of pro-insulin were present in the serum of a patient with an islet-cell tumour and in the urine of an untreated juvenile diabetic.[26] There may be a relationship between these observations and the high but seemingly biologically ineffective levels of "immunoreactive insulin" that have often been observed in the early stages of human diabetes.

VI. Conclusions and Prospects

The object of this essay has been to give a perspective view of the secretory cycle of the β cell of the Islets of Langerhans. Due to the stimulus provided by the discovery of proinsulin in 1967, our knowledge of this intracellular sequence of events has blossomed in the last few years, and the full consequences have yet to be realized. As an object of study in protein chemistry, the chemical synthesis of proinsulin may open the way to some comprehension of the directing role of the connect-ing peptide in the formation of native proinsulin from the fully reduced and unfolded form and as a potential source of synthetic insulin and its analogues for use in clinical medicine. Larger polypeptides such as ribonuclease have already been synthesized by solid-phase methods and an additional supply of insulin may be of importance to a world popula-tion with an ever-increasing number of diabetics. Another obvious possibility with some relevance to human medicine is the study of genetically determined or random defects either in the structure of the connecting peptide or in the enzymes in the β cell that are concerned with the conversion of proinsulin to the active hormone.

Integrated biochemical and cytological studies on many cell types have identified some interrelationships of the nuclear envelope, the endoplasmic reticulum and the Golgi apparatus (collectively termed as the endoplasmic region of the cell[47]) and their role in the biosynthesis, transport and packaging of a wide variety of intracellular materials and other products destined for secretion into the external environment of the cell. In contrast, little is known about the mechanisms involved in the intracellular transport of the products of biosynthesis contained in droplets, vesicles and membrane-limited secretory granules in the exoplasmic space of the cell.[67, 68] In this context, the suggested role of microtubules in the translocation of secretory granules in β cells[42] is worthy of further study.

4

The process of extrusion of the insulin granule from the β cell is generally assumed to involve the fusion of the membrane of the secretory granule with that of the cytoplasmic membrane so that there is an addition of a new area of membrane to that of the cell surface. A similar fusion process of membrane-limited zymogen granules occurs in pancreatic exocrine cells, but the surface area of this cell is not greatly increased during an active phase of secretion[69] and it must be assumed that the exocrine cell, at least, can reutilize excess cell membrane by unidentified mechanisms.

The concept that all the insulin produced by the β cell must pass through the secretory granule stage is an interesting area of disagreement. There is indirect evidence for the secretion of both "newly synthesized insulin", and "stored insulin" from secretory granules, by intact β cells.[70-72] The nature of this possible alternative pathway of secretion is not known and it is not certain whether the "newly synthesized insulin" is insulin or proinsulin. It may well prove to be proinsulin since it has been found[65] that after ingestion of glucose by human subjects there is an initial secretion of insulin, but in the period 60–120 min some 40% of the secretion can be accounted for as "big" insulin (see Section V). In this context, the possibility must be considered that the glucose-induced stimulation of an asynchronous population of β cells could result in the secretion of insulin from a cell (in a storage phase) that contains secretory granules but little endoplasmic reticulum, as well as the premature release of insulin or proinsulin from a β cell in an active stage of biosynthesis and containing few mature secretory granules. In any event, this essay serves to illustrate that at present we are merely at the threshold of knowledge with regard to the regulation of insulin formation and secretion by the pancreatic β cell.

REFERENCES

1. Ottesen, M. (1967). Induction of biological activity by limited proteolysis. *A. Rev. Biochem.* **36**, 55–76.
2. Schachter, M. (Ed.) (1960). *Polypeptides which affect smooth muscles and blood vessels*. Oxford: Pergamon Press.
3. Sachs, H., Fawcett, P., Takabatake, Y. & Portanova, R. (1969). Biosynthesis and release of vasopressin and neurophysin. *Recent Prog. Horm. Res.* **25**, 447–491.
4. Haist, R. E. (1968). Current concepts in diabetes. *J. Can. med. Ass.* **99**, 1017–1025.
5. Lacy, P. E. & Kostianovsky, M. (1967). Method for the isolation of intact islets of Langerhans from the rat pancreas. *Diabetes*, **16**, 35–39.
6. Dixit, P. K., Lowe, I. & Lazarow, A. (1962). Effect of alloxan on the insulin content of micro-dissected mammalian pancreatic islets. *Nature, Lond.* **195**, 358–389.

7. Lacy, P. E. & Williamson, J. R. (1962). Quantitative histochemistry of the islets of Langerhans: II, Insulin content of dissected beta cells. *Diabetes*, **11**, 101–104.

8. Reid, K. B. M., Grant, P. T. & Youngson, A. (1968). The sequence of amino acids in insulin isolated from islet tissue of the cod (*Gadus callarias*). *Biochem. J.* **110**, 289–296.

9. Carpenter, F. H. (1966). Relationship of structure to biological activity of insulin as revealed by degradative studies. *Am. J. Med.* **40**, 750–758.

10. Hunt, L. T. (1969). Hormones, kinins and toxins. In *Atlas of Protein Sequence and Structure*, Vol. 4, pp. D.147–D.172. Ed. by Dayhoff, M. O. Silver Spring, Md.: National Biomedical Research Foundation.

11. Humbel, R. E. (1965). Biosynthesis of the two chains of insulin. *Proc. natn. Acad. Sci. U.S.A.* **53**, 853–859.

12. Grant, P. T. & Reid, K. B. M. (1968). Biosynthesis of an insulin precursor by islet tissue of cod (*Gadus callarias*). *Biochem. J.* **110**, 281–288.

13. Steiner, D. F., Cunningham, D., Spigelman, L. & Aten, B. (1967). Insulin biosynthesis: evidence for a precursor. *Science, N.Y.* **157**, 697–700.

14. Askonas, B. A. & Williamson, A. R. (1967). Biosynthesis and assembly of immunoglobulin G. *Cold Spr. Harb. Symp. quant. Biol.* **32**, 249–254.

15. Klostermeyer, H. & Humbel, R. E. (1966). The chemistry and biochemistry of insulin. *Angew. Chem. Int. Ed. Engl.* **5**, 807–822.

16. Varandani, P. T. (1967). Acceleration of regeneration of insulin activity from its inactive reduced A and B chains by pancreatic glutathione-insulin transhydrogenase. *Biochim. biophys. Acta*, **132**, 10–14.

17. Givol, D., DeLorenzo, F., Goldberger, R. F. & Anfinsen, C. B. (1965). Disulfide interchange and the three-dimensional structure of proteins. *Proc. natn. Acad. Sci. U.S.A.* **53**, 676–684.

18. Trakatellis, A. C. & Schwartz, G. P. (1970). Biosynthesis of insulin in anglerfish islets. *Nature, Lond.* **225**, 548–549.

19. Dintzis, H. M. (1961). Assembly of the peptide chains of haemoglobin. *Proc. natn. Acad. Sci. U.S.A.* **47**, 227–261.

20. Weisblum, B., Cherayil, J. D., Bock, R. M. & Söll, D. (1967). An analysis of arginine codon multiplicity in rabbit hemoglobin. *J. molec. Biol.* **28**, 275–280.

21. Steiner, D. F. & Oyer, P. E. (1967). The biosynthesis of insulin and a probable precursor of insulin by a human islet cell adenoma. *Proc. natn. Acad. Sci. U.S.A.* **57**, 473–480.

22. Tung, A. K. & Yip, C. C. (1968). The biosynthesis of insulin and 'proinsulin' in foetal bovine pancreas. *Diabetologia*, **4**, 68–70.

23. Chang, A. Y. (1970). Insulin synthesis and secretion by isolated islets of spontaneously diabetic Chinese hamsters. In *Structure and Metabolism of the Pancreatic Islets*, pp. 515–526. Ed. by Falkmer, S., Hellman, B. & Täljedal, I. B. Oxford: Pergamon Press.

24. Steiner, D. F. (1967). Evidence for a precursor in the biosynthesis of insulin *Trans. N.Y. Acad. Sci.* **30**, 60–68.

25. Chance, R. E., Ellis, R. M. & Bromer, W. W. (1968). Porcine proinsulin: characterization and amino acid sequence. *Science, N.Y.* **161**, 165–167.

26. Steiner, D. F., Clark, J. L., Nolan, C., Rubenstein, A. H., Margoliash, E., Aten, B. & Oyer, P. E. (1969). Proinsulin and the biosynthesis of insulin. *Recent Prog. Horm. Res.* **25**, 207–282.

27. Grant. P. T. (Unpublished results).
28. Steiner, D. F. & Clark, J. L. (1968). The spontaneous reoxidation of reduced beef and rat proinsulins. *Proc. natn. Acad. Sci. U.S.A.* **60**, 622–629.
29. Wilson, S., Aprile, M. A. & Sasaki, L. (1967). The antigenic loci of insulin. *Can. J. Biochem.* **45**, 1135–1144.
30. Rubenstein, A. H., Milani, F., Pilkis, S. & Steiner, D. F. (1969). Proinsulin: secretion, metabolism, immunological and biological properties. *Postgrad. Med. J.* **45**, Suppl. 476–481.
31. Rubenstein, A. H., Clark, J. L., Milani, F. & Steiner, D. F. (1969). Secretion of proinsulin C-peptide by pancreatic β cells and its circulation in blood. *Nature, Lond.* **224**, 697–699.
32. Yip, C. C. & Logothetopoulos, J. (1969). A specific anti-proinsulin serum and the presence of proinsulin in calf serum. *Proc. natn. Acad. Sci. U.S.A.* **62**, 415–419.
33. Shaw, W. N. & Chance, R. E. (1968). Effect of porcine proinsulin *in vitro* on adipose tissue and diaphragm of the normal rat. *Diabetes*, **17**, 737–745.
34. Chance, R. E. (1969). Discussion following article by Steiner, D. F., Clark, J. L., Nolan, C., Rubenstein, A. H., Margoliash, E., Aten, B. & Oyer, P. B. (1969). Proinsulin and the biosynthesis of insulin. *Recent Prog. Horm. Res.* **25**, 207–282.
35. Schmidt, D. D. & Arens, A. (1968). Proinsulin vom Rind. Isolierung, Eigenschaften und seine Aktivierung durch Trypsin. *Hoppe-Seyler's Z. physiol. Chem.* **349**, 1157–1168.
36. Plapp, B. V., Raftery, M. A. & Cole, R. D. (1967). The tryptic digestion of *S*-aminoethylated ribonuclease. *J. biol. Chem.* **242**, 265–270.
37. Canfield, R. E. (1963). The amino acid sequence of egg white lysozyme. *J. biol. Chem.* **238**, 2698–2707.
38. Schramm, M. (1967). Secretion of enzymes and other macromolecules. *A. Rev. Biochem.* **36**, 307–320.
39. Fawcett, D. W., Long, J. A. & Jones, A. L. (1969). The ultrastructure of endocrine glands. *Recent Prog. Horm. Res.* **25**, 315–380.
40. Howell, S. L., Kostianovsky, M. & Lacy, P. E. (1969). Beta granule formation in isolated islets of Langerhans. *J. Cell Biol.* **42**, 695–705.
41. Greider, M. H., Howell, S. L. & Lacy, P. E. (1969). Isolation and properties of secretory granules from rat islets of Langerhans. II: Ultrastructure of the beta granule. *J. Cell Biol.* **41**, 162–166.
42. Lacy, P. E., Howell, S. L., Young, D. A. & Fink, C. J. (1968). A new hypothesis of insulin secretion. *Nature, Lond.* **219**, 1177–1179.
43. Newcomb, E. H. (1969). Plant microtubules. *A. Rev. Pl. Physiol.* **20**, 253–288.
44. Schmitt, F. O. (1968). Fibrous proteins—neuronal organelles. *Proc. natn. Acad. Sci. U.S.A.* **60**, 1092–1101.
45. Grant, P. T., Reid, K. B. M., Coombs, T. L., Youngson, A. & Thomas, N. W. (1970). Distribution and radioactivity of acid–ethanol soluble protein, proinsulin and insulin in subcellular fractions from islet tissue of the cod (*Gadus callarias*). In *Structure and Metabolism of the Pancreatic Islets*, pp. 349–361. Ed. by Falkmer, S., Hellman, B. & Täljedal, I. B. Oxford: Pergamon Press.
46. Sorenson, R. L., Steffes, M. W. & Lindall, A. W. (1970). Sub-cellular localization of proinsulin to insulin conversion in isolated rat islets. *Endocrinology*, **86**, 88–96.

47. de Duve, C. (1969). The lysosome in retrospect. In *Lysosomes in Biology and Pathology*, Vol. 1, pp. 3–40. Ed. by Dingle, J. T. & Fell, H. B. London: North Holland.

48. Coffey, J. W. & de Duve, C. (1968). Digestive activity of lysosomes. I: The digestion of proteins by extracts of rat liver lysosomes. *J. biol. Chem.* **243**, 3255–3263.

49. Grant, P. T., Coombs, T. L., Thomas, N. W. & Sargent, J. R. (1970). The conversion of [^{14}C]proinsulin to insulin in isolated sub-cellular fractions of fish islet preparations. In *Sub-cellular Organization and Function in Endocrine Tissues*. Ed. by Heller, H. & Lederis, K. Cambridge University Press (to be published).

50. Rutter, W. J., Kemp, J. D., Bradshaw, W. S., Clark, W. R., Ronzio, R. A., & Sanders, T. G. (1968). Regulation of specific protein synthesis in cyto-differentiation. *J. cell. Physiol.* **72**, Supp. 1, 1–18.

51. Orci, L., Rufener, C., Pictet, R., Renold, A. E. & Rouiller, Ch. (1970). Present state of the evidence for mixed endocrine and exocrine pancreatic cells in spiny mice. In *Structure and Metabolism of the Pancreatic Islets*, pp. 37–52. Ed. by Falkmer, S., Hellman, B. & Täljedal, I. B. Oxford: Pergamon Press.

52. Logothetopoulos, J., Kaneko, M., Wrenshall, G. A. & Best, C. H. (1964). Zinc granulation and extractable insulin in islet cells following hyperglycemia or prolonged insulin treatment. In *The Structure and Metabolism of the Pancreatic Islets*, pp. 251–265. Ed. by Brolin, S. E., Hellman, B. & Knutson, H. Oxford: Pergamon Press.

53. Adams, M. J., Blundell, T. L., Dodson, E. J., Dodson, G. G., Vijayan, M., Baker, E. N., Harding, M. M., Hodgkin, D. C., Rimmer, B. & Sheat, S. (1969). Structure of rhombohedral 2 Zinc insulin crystals. *Nature, Lond.* **224**, 491–495.

54. Coombs, T. L. & Grant, P. T. (Unpublished results).

55. Butler, J. A. V., Dodds, E. C., Phillips, D. M. P. & Stephen, J. M. L. (1948). The action of chymotrypsin and trypsin on insulin. *Biochem. J.* **42**, 116–122.

56. Slobin, L. I. & Carpenter, F. H. (1966). Kinetic studies on the action of carboxypeptidase A on bovine insulin and related model peptides. *Biochemistry, Easton*, **5**, 499–508.

57. Smith, E. L., Hill, R. L. & Borman, A. (1958). Activity of insulin degraded by leucine amino peptidase. *Biochim. biophys. Acta*, **29**, 207–208.

58. Taniuchi, H., Morávek, L. & Anfinsen, C. B. (1969). Ligand-induced resistance of staphylococcal nuclease and nuclease-T to proteolysis by subtilisin, α-chymotrypsin and thermolysin. *J. biol. Chem.* **244**, 4600–4605.

59. Mayhew, D. A., Wright, P. H. & Ashmore, J. (1969). Regulation of insulin secretion. *Pharmac. Rev.* **21**, 183–212.

60. Behrens, O. K. & Grinnan, E. L. (1969). Polypeptide hormones. *A. Rev. Biochem.* **38**, 83–112.

61. Hales, C. N. (1967). Some actions of hormones in the regulation of glucose metabolism. In *Essays in Biochemistry*, Vol. 3, pp. 73–104. Ed. by Campbell, P. N. & Greville, G. D. London: Academic Press.

62. Coore, H. G. & Randle, P. J. (1964). Regulation of insulin secretion studied with pieces of rabbit pancreas incubated *in vitro*. *Biochem. J.* **93**, 66–78.

63. Kinash, B. & Haist, R. E. (1954). Continuous intravenous infusion in the rat, and the effect on the islets of Langerhans of the continuous infusion of glucose. *Can. J. Biochem.* **32**, 428–433.

64. Lin, B. J. & Haist, R. E. (1969). Insulin biosynthesis: effects of carbohydrates and related compounds. *Can. J. Physiol. Pharm.* **47**, 791–801.
65. Gorden, P. & Roth, J. (1969). Circulating insulins. "Big" and "Little". *Archs intern. Med.* **123**, 237–247.
66. Rubenstein, A. H., Cho, S. & Steiner, D. F. (1968). Evidence for proinsulin in human urine and serum. *Lancet*, **i**, 1353–1355.
67. Jamieson, J. D. & Palade, G. E. (1968). Intracellular transport of secretory proteins in the pancreatic exocrine cell. III: Dissociation of intracellular transport from protein synthesis. *J. Cell Biol.* **39**, 580–588.
68. Jamieson, J. D. & Palade, G. E. (1968). Intracellular transport of secretory proteins in the pancreatic exocrine cell. IV: Metabolic requirements. *J. Cell Biol.* **39**, 589–603.
69. Ichikawa, A. (1965). Fine structural changes in response to hormonal stimulation of the perfused canine pancreas. *J. Cell Biol.* **24**, 369–385.
70. Grodsky, G., Landahl, H., Curry, D. & Bennett, L. (1970). *In vitro* studies suggesting a two-compartment model for insulin secretion. In *Structure and Metabolism of the Pancreatic Islets*, pp. 409–421. Ed. by Falkmer, S., Hellman, B. & Täljedal, I. B. Oxford: Pergamon Press.
71. Creutzfeldt, W., Creutzfeldt, C. & Frerichs, H. (1970). Evidence for different modes of insulin secretion. In *Structure and Metabolism of the Pancreatic Islets*, pp. 181–198. Ed. by Falkmer, S., Hellman, B. & Täljedal, I. B. Oxford: Pergamon Press.
72. Cerasi, E. & Luft, R. (1970). The pancreatic β-cells in the pathogenesis of human *diabetes mellitus*. In *Structure and Metabolism of the Pancreatic Islets*, pp. 533–543. Ed. by Falkmer, S., Hellman, B. & Täljedal, I. B. Oxford: Pergamon Press.
73. Creutzfeldt, W., Creutzfeldt, C., Frerichs, H., Perings, E. & Sickinger, K. (1969). The morphological substrate of the inhibition of insulin secretion by diazoxide. *Horm. Metab. Res.* **1**, 53–64.
74. Janoff, A. & Zeligs, J. D. (1968). Vascular injury and lysis of basement membrane *in vitro* by a neutral protease of human leukocyctes. *Science, N.Y.* **161**, 702–704.
75. Tesar, J. T., Koenig, H. & Hughes, C. (1969). Hormone storage granules in the beef anterior pituitary. I. Isolation, ultrastructure and some biochemical properties. *J. Cell Biol.* **40**, 225–235.
76. Hodgkin, D. C. Personal communication.

Lactose Synthetase: Evolutionary Origins, Structure and Control

K. BREW

*Department of Biochemistry, University of Leeds,
Leeds LS2 9LS, England*

I. Introduction

Evidence from comparative morphology and embryology indicates that the mammary gland is derived from sebaceous glands of the skin.[1] Although mammary function is a major defining characteristic of both the Therian and Monotreme mammals, the most recent common ancestor of these two mammalian groups was probably a Therapsid reptile.[2] However, there can be little doubt that the evolutionary origin of the mammary gland is not much more remote than that of the mammals. Evidently, in some not too remote biological era, the ability to synthesise and secrete milk arose in some ancestor of the mammals through a series of mutational events which resulted in the modification of pre-existing capacities in an area of skin. On examination of the biochemical activities of modern mammary glands, these are found to consist of the ability to synthesize and secrete a group of milk-specific proteins, fat, and the disaccharide lactose, under the regulation of a system of hormonal and neural controls.

Apart from its occurrence in the mammary gland and milk, lactose has been found only in a few plant tissues, and the ability to synthesize and secrete lactose, especially in the large quantities found in milk, is a unique characteristic of the lactating mammary gland. Well-defined examples of biochemical functions of relatively recent origin are rare, especially cases where we can define the origins of the proteins which control the activities.[3] The justification for confining the subject of this essay to one enzyme system, lactose synthetase,* is that at present this is unique in providing information not only about the origins of a biochemical activity and some of its control processes but also in that it illuminates some interesting facets of the control of enzyme and organ specificity.

The terminology, mechanisms and nature of evolutionary change in the structures of proteins has been previously comprehensively reviewed in this series by Dixon[3] and will not be discussed in any detail here.

II. Structure and Function

A. THE INTERMEDIATES AND ENZYMES OF LACTOSE BIOSYNTHESIS

The accepted pathway of lactose synthesis in the lactating mammary gland is summarized in Fig. 1. It is not proposed to discuss the evidence for this scheme which has been dealt with elsewhere (for example, Refs. 4 and 45). All the steps and enzymes of the pathway are common to several synthetic pathways which utilize UDP-galactose as a precursor.[5] These include galactolipid synthesizing systems[6] and specific galactosyl transferases involved in the synthesis of glycoproteins,[7, 16] and blood-group substances.[8] Only the final step of the pathway is unique for lactose biosynthesis. In this the galactosyl moiety of UDP-galactose is transferred to the 4-OH group of glucose to produce lactose (cf. Fig. 6, II):

$$\text{UDP-galactose} + \text{glucose} \rightarrow \text{lactose} + \text{UDP} \tag{1}$$

---

* The more correct terminology would be uridine diphosphate-D-galactose: D-glucose-β-4-galactosyl transferase (though EC 2.4.1.22 gives UDPgalactose: D-glucose 1-galactosyltransferase). For reasons of space and simplicity, the trivial name lactose synthetase will be used throughout this essay. N-acetyllactosamine synthetase will similarly be used to denote uridine diphosphate-D-galactose: N-acetyl-D-glucosamine-β-4-galactosyl transferase.

Abbreviation: NAL = N-acetyllactosamine.

An enzyme system which catalyzes this reaction was first isolated from homogenates of lactating guinea-pig and bovine mammary glands[9] where it was associated with a particulate fraction of the homogenate. Later, a soluble form of the enzyme was found to be present in milk.[10] Most of the work to be discussed here has been carried out on the soluble milk enzyme, for the obvious reasons of easy availability and of avoiding the problems associated with trying to solubilize a particulate enzyme. Recently, interest has revived concerning the nature of the particulate enzyme. For the purposes of this discussion, it is simpler to discuss initially the nature and properties of the soluble milk enzyme, and later to relate them to its subcellular organization and to the properties of the particulate enzyme.

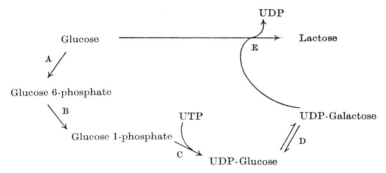

FIG. 1. The intermediates and enzymes of the lactose biosynthetic pathway in the lactating mammary gland. The enzymes are denoted as follows: A, hexokinase; B, phosphoglucomutase; C, UDP-glucose pyrophosphorylase; D, UDP-galactose-4-epimerase; E, lactose synthetase.

B. THE A AND B PROTEINS OF LACTOSE SYNTHETASE

Milk lactose synthetase appeared initially to be labile to most purification procedures[10] until Brodbeck and Ebner[11] discovered that a partially purified fraction containing the enzyme was resolved into two components by gel filtration. Neither component *per se* was active for lactose synthesis, but on combining them, activity was regenerated. The components were designated the A and B proteins of lactose synthetase, and were suggested to be subunits of the enzyme.[11] The component of lower molecular weight, the B protein, was later found by the same workers to be identical with a well-characterized milk whey protein, α-lactalbumin.[12] As α-lactalbumin is a major protein component of milk whey, and lactose synthetase might be expected to be active within the lactating mammary epithelial cells, this was an unexpected result. Now that more is known about lactose synthetase, the explanation of this has become clear, and will be discussed later in this essay.

C. ROLE OF THE A AND B PROTEINS

Further examination of the substrate specificity of lactose synthetase and its component proteins by Brew, Vanaman and Hill[13] revealed that although neither the A nor B proteins was active for lactose synthesis in isolation, the A protein catalysed another reaction [reaction (2)].

$$\text{UDP-D-galactose} + N\text{-acetyl-D-glucosamine} \rightarrow \text{UDP} + N\text{-acetyllactosamine} \quad (2)$$

In the presence of increasing concentrations of the B protein, reaction (2) is inhibited and reaction (1) is enhanced (see Fig. 2). The net effect of the B protein is to change the substrate (acceptor) specificity of the system from N-acetylglucosamine to glucose. Within the cell this would amount to diverting the utilization of the nucleotide sugar (UDP-galactose) from the synthesis of one product to another. This effect of α-lactalbumin on the specificity of N-acetyllactosamine synthetase is an apparently unique activity for a protein, and because of this it was designated the "specifier protein" of lactose synthetase.[13]

Figure 3 shows the effect of increasing specifier protein concentrations on the activity of two different NAL synthetase concentrations for lactose synthesis. It is apparent that a simple titration of the two components with each other does not occur, as would be expected under normal circumstances for the reassembly of two subunits of an enzyme system. For this reason it is slightly misleading to speak of the A and B proteins as subunits of lactose synthetase, as this has led some workers to assume that at some point there is an "excess" of the A or B proteins in a system, and that the A and B proteins can be separately assayed by assuming that the generation of lactose synthetase activity results from a titration of one subunit with the other. This assumption can give rise to misleading results (see Section IIIC).

Kinetically, it is possible to use the data for lactose synthetase activity with different B protein concentrations in order to obtain an apparent binding constant for the B protein in the enzyme system, although at high B protein concentrations inhibition of lactose synthetase activity is observed (see Figs. 2 and 10). It is apparent that the B protein is binding either to the A protein or to some intermediate in the enzyme catalysed reaction. That the effect of the B protein on the overall specificity of the enzyme system is not the result of two successive reactions catalysed separately by the A and B proteins is clearly shown

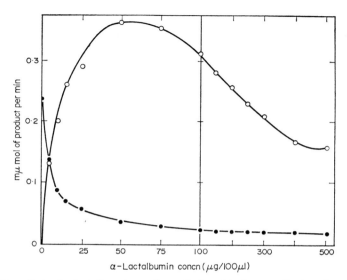

FIG. 2. The effect of α-lactalbumin on the activity of lactose synthetase A protein for lactose and N-acetyllactosamine synthesis (Hill, Brew, Vanaman, Trayer and Mattock[18]): ○, Lactose synthesis; ●, N-acetyllactosamine synthesis.

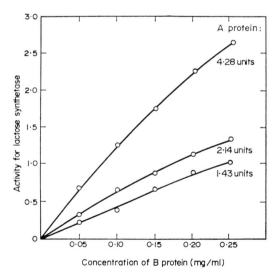

FIG. 3. The effect of α-lactalbumin concentration on lactose synthetase activity with three different concentrations of A protein.

by the fact that lactose synthesis does not occur when the A and B proteins are separated by a dialysis membrane.[13]

D. OTHER ENZYME SYSTEMS OF CHANGEABLE SPECIFICITY

Only a very small group of enzymes exhibits changes in specificity under different conditions. All of the present known examples, apart from lactose synthetase, are of bacterial origin and none is really analogous with it, nor are they comparable among themselves.

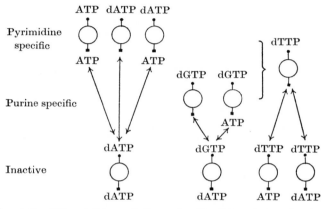

Fig. 4. Scheme for different forms of ribonucleoside diphosphate reductase. The ring and square denote separate binding sites which affect the specificity and level of activity of the enzyme respectively. Thus differences in nucleoside triphosphate binding to the first type of site changes the enzyme from pyrimidine-specific to purine-specific or non-specific forms (left to right in the diagram). Binding to the second type of site results in the interconversion of the active and inactive forms (denoted by arrows in the diagram). Taken from Brown and Reichard.[15]

For example, tryptophan synthetase from *E. coli* consists of two easily separable components which catalyse successive steps in the biosynthesis of tryptophan. On combination, a reaction which is the sum of the two reactions is catalysed at a far higher rate and without the formation of a free intermediate.[14] A very different and particularly interesting example is ribonucleoside diphosphate reductase.[15] This again is separable into two dissimilar components which are inactive and regenerate activity on combination. This enzyme, however, exhibits changes in specificity and activity on the addition of nucleoside triphosphates. Two binding sites appear to be present on one of the subunits of the enzyme. One site modulates the specificity of the enzyme (for purines or pyrimidines or both, depending on the nature of the bound effector) and the other controls the activity (i.e. activity when ATP is bound or inactivity when dATP is bound). The complexity of forms of the enzyme is shown in Fig. 4.

As in tryptophan synthetase, the specificity change in lactose synthetase is mediated by combination of the two protein components rather than by the allosteric effects produced by the binding of small molecules, as in ribonucleoside diphosphate reductase. A major difference does however exist in the metabolic role of the specificity change. Thus, the activity of one component of lactose synthetase (the A protein) in isolation has a distinct biological function from that of the complete enzyme system, whereas in tryptophan synthetase all the reactions are concerned with the same biosynthetic pathway.

E. BIOLOGICAL SIGNIFICANCE OF N-ACETYLLACTOSAMINE SYNTHETASES

Although small quantities of N-acetyllactosamine are found in the milk of some mammals, where the substance appears to act as a growth factor for specific intestinal flora, the biological importance of enzymes active in its synthesis would not appear to be very great. Yet such enzymes were studied by Roseman and his group some time before the discovery of the activity in the A protein of lactose synthetase. The interest in this case was in N-acetyllactosamine synthetase as a component of the system for the biosynthesis of exported glycoproteins in mammalian tissues.[16] An important terminal trisaccharide, which is found in the carbohydrate moiety of several secreted glycoproteins including fetuin and orosomucoid, is shown here:

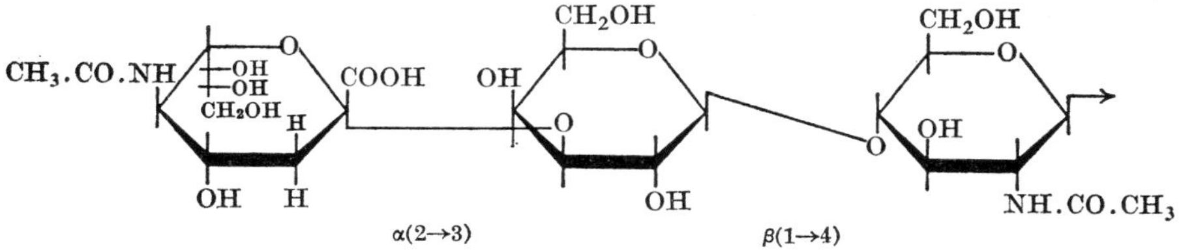

N-Acetylneuraminic acid   D-Galactose N-Acetyl-D-glucosamine

In the biosynthesis of the prosthetic group, the monosaccharide units are thought to be added successively on to the apoprotein after its biosynthesis on the rough endoplasmic reticulum by a series of glycosyl transferases each specific for a nucleotide sugar (UDP derivatives, apart from CMP-sialic acid), and for a monosaccharide acceptor.[7] In this scheme, only one transferase need be specific for an amino acid side chain of the glycoprotein, that is, the enzyme which adds the first monosaccharide unit to the polypeptide chain and which determines the point of attachment of the prosthetic group.[7, 17] Analogous enzymes appear to operate in the biosynthesis of the blood group substances (see Ref. 8). It should be noted that other galactosyl transferases besides N-acetyllactosamine synthetase exist, which transfer the galactose as the α anomer or to a different acceptor, or to a different OH group in the same acceptor.[7, 17]

Enzymes active for the biosynthesis of N-acetyllactosamine and also for the transfer of galactose to orosomucoid treated to remove its terminal sialic acid and galactose residues (called glycoprotein I) were found in a variety of tissues including the lactating mammary gland.[16] Hill and co-workers found that the A protein of lactose synthetase will also catalyse the transfer of galactose to this glycoprotein acceptor.[18] Furthermore, the N-acetyllactosamine synthetase of liver which cannot normally be active in lactose synthesis, is affected in exactly the same way by bovine milk B protein as was bovine A protein, i.e. N-acetyl-lactosamine synthesis was inhibited and lactose synthesis generated.[13] Perhaps the most surprising finding is that N-acetyllactosamine synthetase in chick embryo cerebrospinal fluid and chick embryo brain homogenates is also turned over to lactose synthesis by bovine B protein.[18] The chicken of course does not normally synthesize lactose, and does not have an evolutionary ancestor that synthesized lactose. The obvious conclusion is that the activity of lactose synthetase A protein is an intrinsic property of all N-acetyllactosamine synthetases regardless of their origin. This is an important feature when we come to consider the mechanism of lactose synthesis. Lactose synthesis is therefore in effect accomplished by modification of a piece of glyco-protein-synthesizing machinery in the mammary gland through the mediation of the mammary gland-specific protein α-lactalbumin.

F. STRUCTURE AND FUNCTION OF α-LACTALBUMIN

1. Primary Structure

Having been first purified in 1939,[19] α-lactalbumin was well charac-terized in physical and chemical terms in the 26 years before the discovery of its biological activity (see Ref. 20 for a review). The general chemical characteristics had also been found to show a general resem-blance to those of the enzyme lysozyme,[21] and when the amino acid sequence of bovine α-lactalbumin was determined by Brew, Vanaman and Hill[22] a pronounced structural similarity to lysozyme was revealed.

Homology* between the two proteins is clearly revealed by aligning the primary structures in the manner shown in Fig. 5. The insertion of four gaps, including one of three residues, and one of two residues in the α-lactalbumin sequence and one gap of one residue in the lysozyme sequence is required for maximum homology. Although the structural and genetic relationships of the proteins are quite clear, the insertion of gaps in a sequence comparison of this type is a matter that requires some care. Other authors have removed the only gap in the lysozyme sequence (position 128 in the comparison), as this has little value in establishing the relationship.[23] This rearrangement removes the homology between the eighth ½ cystinyl residues in the sequences. The α-lactalbumin and lysozyme structures have obviously arisen by a mechanism of duplication of an ancestral gene, followed by separate divergent evolution of the products, as evolutionary convergence can only be argued for cases where proteins are functionally similar. Therefore, lack of homology in a cysteinyl residue would necessitate a pathway of evolutionary change in one case that passed through an intermediate form with an unpaired cysteinyl residue. Such an intermediate might be expected to differ considerably in its conformational stability, and is therefore improbable. One possible satisfactory pathway could, however, be through an inversion of residues 127 and 126 (see Ref. 3).

The similarity between α-lactalbumin and lysozyme is about the same as between the α and β chains of haemoglobin (about 44% of residues being identical in both cases) and between the zymogens of the pancreatic proteases (see Ref. 3). The obvious significance of this case is that the biological functions and apparent biochemical activities of α-lactalbumin and lysozyme are very different.

2. *The Common Genetic Ancestor of α-Lactalbumin and Lysozyme*

Lactose and lactose synthetase specifier protein are unique mammalian products, but lysozyme is found both in mammalian tissues and secretions and in avian egg-whites.[24] Analogous, but apparently not homologous, lysozymes are also found in a wide range of species, e.g. papaya latex[25] and T-even phages.[26] Although the most recent common ancestor of α-lactalbumin and lysozyme may have had a different, possibly intermediate, type of activity, the hypothesis that the ancestral protein was a lysozyme[22] deserves some attention. The presence of many deletions in the α-lactalbumin sequence in Fig. 5, but one or even none in the lysozyme sequence, suggests that α-lactalbumin may have

---

* Homology is used in this essay in its strictly biological sense for structures whose controlling genes have a common ancestor, and is implied from the similarities in amino acid sequences (see Ref. 3).

```
                1                                  10
α-Lactalbumin   Glu-Gln-Leu-Thr-Lys-CYS-GLU-Val-Phe-ARG-Glu-LEU-LYS         Asp-LEU-Lys-GLY-TYR-Gly-GLY (20)
Lysozyme
  Chicken       1                                       10                                  20
                Lys-Val-Phe-Gly-Arg-CYS-GLU-Leu-Ala-Ala-Ala-Met-LYS-Arg-His-Gly-LEU-Asp-Asn-TYR-Arg-GLY
  Human         1                                  10                                       20
                Lys-Val-Phe-Glu-Arg-CYS-GLU-Leu-Ala-ARG-Thr-LEU-LYS-Arg-Leu-Gly-Met-Asp-GLY-TYR-Arg-GLY

                                                   30                                       40
α-Lactalbumin   Val-SER-LEU-Pro-Glu-TRP-VAL-CYS-Thr-Thr-PHE-His-Thr-SER-GLY-TYR-Asp-THR-Glu-ALA-Ile-Val
Lysozyme                                    30
  Chicken       Tyr-SER-LEU-Gly-Asn-TRP-VAL-CYS-Ala-Ala-Lys-PHE-Glu-SER-Asn-Phe-Asn-THR-Gln-ALA-Thr-Asn
                                            30                                  40
  Human         Ile-SER-LEU-Ala-Asn-TRP-Met-CYS-Leu-Ala-Lys-Trp-Glu-SER-GLY-TYR-Asn-THR-Arg-ALA-Thr-Asn

                                                   50                                       60
α-Lactalbumin   Glu-ASN         Asn-Gln-SER-THR-ASP-TYR-GLY-Leu-PHE-GLN-ILE-ASN-Asn-Lys-Ile-TRP-CYS-Lys-Asn
Lysozyme                                50                              60
  Chicken       Arg-ASN-Thr     Asp-Gly-SER-THR-ASP-TYR-GLY-Ile-Leu-GLN-ILE-ASN-Ser-Arg-Trp-TRP-CYS-Asn-Asp
                                        50                                                  60
  Human         Tyr-Asn-Ala-Gly-Asp-Arg-SER-THR-ASP-TYR-GLY-Ile-PHE-GLN-ILE-Asn-Ser-Arg-Tyr-TRP-CYS-Asn-Asp

                                        70                                  80
α-Lactalbumin   Asp-Gln-Asp-PRO-His-SER-Ser-ASN-Ile-CYS-ASN-ILE-SER-CYS-Asp-Lys-Phe-LEU-Asn-Asn-ASP-Leu
Lysozyme                    70                                  80
  Chicken       Gly-Arg-Thr-PRO-Gly-SER-Arg-ASN-Leu-CYS-ASN-ILE-Pro-CYS-Ser-Ala-Leu-LEU-Ser-Ser-ASP-Ile
                        80                                  80
  Human         Gly-Lys-Thr-PRO-Gly-Ala Val,Asn Ala-CYS-His-Leu-SER-CYS-Ser-Ala-Leu-LEU-Gln-Asp-Asn-Ile

                                    90                                  100
α-Lactalbumin   THR-Asn-Asn-Ile-Met-CYS-Val-LYS-LYS-ILE-Leu     ASP-Lys-Val-GLY-ILE-ASN-Tyr-TRP-Leu-ALA
Lysozyme            90                                      100                                 110
  Chicken       THR-Ala-Ser-Val-Asn-CYS-Ala-LYS-LYS-ILE-Val-Ser-ASP-Gly-Asp-GLY-Met-ASN-Ala-TRP-Val-ALA
                90                              100                                             110
  Human         Ala-Asp-Ala-Val-Ala-CYS-Ala-LYS-Arg-Val-Arg     ASP-Pro-Gln-GLY-ILE-Arg-Ala-TRP-Val-ALA

                            110                                                     120         123
α-Lactalbumin   His-Lys-Ala-Leu-CYS-Ser-Glu-Lys-Leu-Asp-GLN-Trp-Leu                 CYS-Glu-Lys-LEU
Lysozyme                                            120                                         129
  Chicken       Trp-Arg-Asn-Arg-CYS-Lys-Gly-Thr-Asp-Val-GLN-Ala-Trp-Ile-Arg-Gly-CYS         Arg-LEU
                                                    120                                         129
  Human         Trp-Arg-Asn-Arg-CYS-Gln-Asn-Arg-Asp-Val-Arg-Gln-Tyr-Val-Gln-Gly-CYS         Gly-Val
```

FIG. 5. A comparison of the amino acid sequences of bovine α-lactalbumin (Brew, Vanaman and Hill[22]) with those of hen egg-white lysozyme (Canfield and Liu;[24a] Jollès[24b]) and human lysozyme (Canfield;[27a] Jollès and Jollès[27b]).

undergone a great deal more evolutionary modification and that the lysozyme structure may be more similar to that of the common ancestor.

Recently, the nearly complete structure of human lysozyme has been published.[27a, b] This sequence is obviously more similar to that of hen's egg-white lysozyme than to bovine α-lactalbumin, but shows a similarity to α-lactalbumin in some positions where hen's egg-white lysozyme differs (also shown in Fig. 5). Some information about the possible relative distance of divergence of this protein from bovine α-lactalbumin, as compared with the distance for hen's egg-white lysozyme–bovine α-lactalbumin can be obtained by summing the minimal base changes in the mRNA codons required to interconvert these pairs of sequences.[3, 23] The result is tabulated here.

Pair of proteins	Minimal base mutational distance
Human lysozyme–bovine α-lactalbumin	111
Hen's egg-white lysozyme–bovine α-lactalbumin	110

This must be regarded with caution, as the comparisons are based on a generous assumption of similarities in regions where the sequence of human lysozyme is uncertain. The provisional conclusions that the most recent common ancestor of α-lactalbumin and lysozyme is more distant than the point of divergence of the mammalian and avian lines is especially doubtful, as the development of a new function in one line (α-lactalbumin) may have had an accelerating effect on the rate of evolutionary change, swamping out the slow constant change generally observed in a functionally conservative line. We must await the completion of many more sequences of α-lactalbumins and lysozymes before drawing any positive conclusions about the antiquity and the nature of the common ancestor, but there is obviously a good chance that eventually we will succeed in this aim.

3. Conformation and Mode of Action

Initially, a gross similarity was noticed between the substrate specificities of lysozyme and α-lactalbumin. Lysozyme catalyzes the hydrolysis of a $\beta(1 \rightarrow 4)$ glycosidic linkage between N-acetylneuraminic acid and N-acetylglucosamine residues in a bacterial cell wall oligosaccharide; lactose synthetase catalyses the synthesis of a $\beta(1 \rightarrow 4)$ linkage between galactose and glucose (see Fig. 6).[22] When it was found that the overall activity of α-lactalbumin was to change the specificity of N-acetyllactosamine synthetase, the activities of α-lactalbumin and lysozyme began to appear very diverse. Before questioning the compatability of this difference with the structural similarity and common ancestry of the two proteins, it is advantageous to examine more closely the similarities and differences in their structures.

A similarity in the three dimensional structures of α-lactalbumin and lysozyme is immediately suggested by their similarities in amino acid sequence. When chemical studies showed that the pairing of the eight pairs of homologous ½ cystinyl residues into disulphide bonds was identical in the two proteins[18] this possibility became very attractive. Conservation of conformation has been observed between other pairs of homologous proteins which differ to some extent in their functions (e.g. chymotrypsinogen and elastase, see Ref. 28), even when, as in the

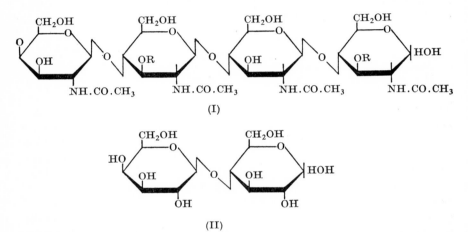

(I)

(II)

Fig. 6. A comparison of the structure of a tetrasaccharide substrate of lysozyme (I) with that of lactose (II). R = —CH(CH₃)COOH.

case of haemoglobin and myoglobin, the amino acid sequences differ far more than those of α-lactalbumin and lysozyme.[3, 29] Model building studies showed that the α-lactalbumin amino acid sequence will fit into the lysozyme conformation previously determined by X-ray crystallography by Phillips and co-workers.[30, 31] The convincing way that the deletions in the α-lactalbumin sequence are accommodated in the structure, and the "complementary" nature of amino acid substitutions in the hydrophobic core region make this model a reasonable basis for further investigations. The optical rotatory dispersion,[32] circular dichroism[33, 34] and nuclear magnetic resonance spectra[34] of bovine α-lactalbumin are largely consistent with the model. There is no doubt that the two proteins differ in detail in their conformations, and the fact that α-lactalbumin and lysozyme do not cross-react immunologically has been taken to suggest that considerable conformation differences exist between the proteins.[35] As α-lactalbumins from different species also do not cross-react immunologically,[36] lack of cross-reactivity between α-lactalbumin and lysozyme is hardly surprising, and has no relevance for the overall conformation of the two proteins. In fact, a comparison[37] of one major antigenic site of lysozyme, residues 64–83, with the corresponding region in bovine α-lactalbumin shows a very large proportion of amino acid substitutions. Changes in such regions would be expected to cause profound changes in the antigenic properties of the protein despite the conservation of overall conformation.

The most interesting area of the proposed α-lactalbumin conformation is that which corresponds to the active site region in lysozyme and is situated in a cleft in the surface of the protein. The differences in this region produced by the amino acid substitution in α-lactalbumin make the structure particularly interesting for this discussion. In lysozyme, the cleft is open at both ends to accommodate a large substrate molecule and the amino acid side chains in this area make many contacts with the oligosaccharide substrate.[38] In the α-lactalbumin structure, one end of the cleft and many sites of substrate interaction are cut off by a new hydrophobic region which appears to be generated in part by the substitution of a tyrosine side chain in α-lactalbumin for an alanine in lysozyme (position 107 in Fig. 5). Evidently α-lactalbumin in this region will be incapable of interacting with large molecules, and as the substrates of lactose synthetase are of low molecular weight, this encourages the view that the cleft region of α-lactalbumin may be involved in the biological activity.

Several of the amino acid side chains which are essential for the catalytic activity of lysozyme are evidently not important in α-lactalbumin. Thus, of the two amino acid side chains in lysozyme which are actually involved in catalysis,[38] glutamic acid 35 and aspartic acid 52, although the latter residue is conserved in α-lactalbumin, glutamic acid 35 is replaced by a residue of histidine or threonine, depending on how the sequences are aligned (Fig. 5). Although the imidazole side chain of histidine has the potential for acting as a proton donor, and could thus function in a similar way to glutamic 35 in the activity of lysozyme, it does not appear to do so in α-lactalbumin. Castellino and Hill have found that carboxymethylation of this residue, and in fact of all the histidines in α-lactalbumin, does not eliminate the lactose synthetase specifier activity of the protein.[39] Further examples are tryptophan 63 which is again important for the activity of lysozyme, but is absent from guinea-pig α-lactalbumin and therefore cannot be essential for lactose synthesis. Tryptophan 26 of bovine α-lactalbumin is also missing from human α-lactalbumin and is therefore not essential for biological activity.[40]

Two further major differences between the proteins involve first the multiple deletions in the carboxyl terminal region of α-lactalbumin (Fig. 5) which suggest that a considerable conformation change has occurred in the region of the disulphide bond between $\frac{1}{2}$ cystines in positions 6 and 120[31] and could be associated with the differences in denaturation properties of α-lactalbumin and lysozyme.[35] Second, a major difference exists between the isoelectric points of α-lactalbumin and lysozyme. Lysozymes are very basic proteins with isoelectric points at about pH 11, whereas α-lactalbumins are rather acidic proteins with isoelectric points at about pH 5·0. The positive charge on lysozyme in its pH activity range appears to be important for the bacteriolytic activity and may be required for a general electrostatic interaction with the bacterial cell wall.[24] The acidic nature of α-lactalbumin may be an adaptive change associated with its interaction with N-acetyllactosamine synthetase which is a slightly basic protein.[18] This major change in isoelectric point is however effected through a relatively small number of substitutions of acidic or neutral residues in α-lactalbumin for basic residues in lysozyme, e.g. residues 1, 45, 14 and 15 in lysozyme (see Fig. 5).

In examining the evolutionary development of lactose synthesis, we encounter two apparently irreconcilable concepts. One is our present idea of proteins evolving as a result of single amino acid replacements through a series of forms differing only slightly in activity, producing through gradual change new activities. The other is the relatively recent divergence of α-lactalbumin and lysozyme and their apparently gross difference in biological function.

One type of reaction mechanism for lactose synthesis can be envisaged which is compatible with both these concepts and is consistent with the type of differences between α-lactalbumin and lysozyme. In this scheme, α-lactalbumin will have an activity of transferring an activated galactosyl intermediate from the A protein–UDP-galactose complex to an acceptor characteristic of α-lactalbumin:glucose. A type of intermediate which is formed in the lysozyme catalysed cleavage of an N-acetylneuraminyl—N-acetylglucosaminyl linkage (a C_1 carbonium ion intermediate) could be commonly bound by the active site regions of both the A and B proteins. This type of activity is entirely compatible with the modification in the cleft region of α-lactalbumin, and would explain many puzzling features of the lactose synthetase system, including the lack of species specificity in the interaction of the A and B proteins.

The apparently very different activities of α-lactalbumin and lysozyme may therefore yet involve basically a very similar reaction mechanism.

III. Control of Lactose Synthetase

A. HORMONAL CONTROL

As lactose synthetase activity in a mixture of A and B proteins increases with the A protein concentration at all B protein concentrations (see e.g. Fig. 3), it is not possible to assay the A and B protein concentration in homogenates of mammary tissue by simple assays for N-acetyl-lactosamine synthetase and for lactose synthetase respectively.

This problem can be circumvented by assaying for the B protein in the presence of a high concentration of exogenous A protein.[41] When measurements are made by assuming an "excess" of A protein to be present in the homogenate and using the lactose synthetase activity as a measure of the B protein level, the results obtained for the B protein level are linked to the levels of both the A and B proteins (implicit in Fig. 3). In fact in this sort of measurement, the "A" and "B" protein levels obtained are always linked (see Ref. 42). Such results do not constitute a reliable independent measure of the A and B proteins and are misleading and partly in disagreement with those discussed here.

1. *Changes in A and B Protein Levels during Mammary Development*

During pregnancy, the mammary gland undergoes preparations for its secretory phase of lactation. During this period, the activity of many enzymes increases including those associated with the biosynthesis of UDP-galactose. Lactose synthetase activity appears to be rate limiting for lactose synthesis during this phase and during lactation.[43] It is particularly interesting therefore that at least in the mouse N-acetyl-lactosamine synthetase activity also increases during pregnancy, whereas lactose synthetase "specifier" activity (α-lactalbumin) remains at a low level, although a slight increase is observed towards the end of pregnancy. Immediately following parturition, however, although the A protein remains at its prelactational level, a rapid rise in the level of total B protein is observed in the tissue (Fig. 7).[41]

This asynchrony in the expression of the A and B proteins during pregnancy and lactation can be related to the dual biological role of the A protein. During pregnancy, glycoprotein synthesis (in which the A protein functions) may occur in the mammary gland, but it is only during lactation that the increase in the level of B protein permits high rates of lactose synthesis. It must be noted that at least some of the post-parturitional rise in the B protein level measured in tissue homogenates may be associated with the high content of milk in even well-washed lactating mammary tissue, as the B protein is found in milk in high concentration, but there is no doubt that α-lactalbumin synthesis increases considerably at the onset of lactation.

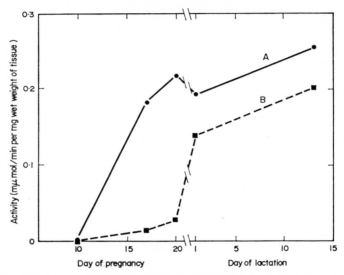

FIG. 7. The levels of the lactose synthetase A and B proteins in homogenates of mammary tissue from mice at different stages of pregnancy and lactation. Fig. from Turkington, Brew, Vanaman & Hill.[41]

2. Control of the Expression of the A and B Proteins

It is natural to inquire as to the basis of the control of the asynchrony of expression of the A and B proteins of lactose synthetase. Topper and co-workers have found that the synthesis of casein and whey proteins can be "induced" in mammary tissue from pregnant mice by incubating the tissue in organ culture in the presence of the hormones insulin, hydrocortisone and prolactin.[44, 45] The effect of the hormones can be divided into two stages: (a) a phase of cell division and differentiation invoked by the action of both insulin and hydrocortisone, and (b) a phase of induction which required the presence of insulin, hydrocortisone and prolactin.[45] Turkington, Brew, Vanaman and Hill[41] measured the levels of the A and B proteins in tissue treated in this way, and found that the increases in the levels of both proteins are obtained only by treatment with the three hormone combination [Fig. 8(a)]. Similarly, the phases of cell division plus differentiation, followed by induction can be observed separately. Differentiation is inhibited by colchicine (an inhibitor of cell division) and induction inhibited by inhibitors of protein synthesis.[41] Probably the most interesting observation is that when the hormone progesterone is included in the system, the increase in the A protein level still occurs, but the B protein remains at a low level. Injection of progesterone into pregnant mice just before parturition also prevents the increase in the level of the B protein. Progesterone has no effect on the "induction" of casein in the organ culture system, and seems to operate specifically in the regulation of α-lactalbumin synthesis.[46]

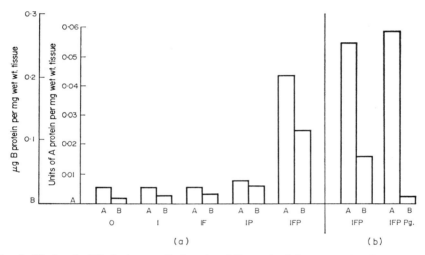

FIG. 8. The levels of the lactose synthetase A and B proteins in homogenates of mammary tissue from pregnant mice after incubation of the tissue in organ culture in the presence of different hormone combinations. The hormones are denoted as follows: O, none; I, insulin; F, hydrocortisone; P, prolactin; Pg, progesterone. (a) Data from Turkington et al.[41]; (b) data from Turkington and Hill.[46]

Progesterone is a hormone which is present during pregnancy, but decreases in level at the onset of lactation, coincident with the appearance of lactose synthesis. It therefore seems that the hormonal regulation of lactose synthetase involves the synergistic and antagonistic interactions of four different hormones and is a very remarkable example of the complexity of the systems which regulate the expression of enzymes and hence the rates of biosynthetic pathways in higher animals.

B. SUBCELLULAR ORGANIZATION OF A AND B PROTEINS

Investigations of the subcellular distribution of the A and B proteins of lactose synthetase have yielded different results which depend on the method used to disrupt the mammary tissue. Using a gentle method of homogenization a particulate fraction containing both components can be prepared.[9, 47] Coffey and Reithel[47] studied the sedimentation properties and enzymic constitution of the particles which appeared similar to the characteristics of the membranes of the Golgi region, and are probably derived from this region of the mammary cells.[47] These particles were interpreted as containing the A and B proteins firmly joined in a stable complex. Similar particles from the mouse mammary gland were suggested to contain not only the A and B proteins, but also a third component which was hypothetically inferred to be required for the formation of a stable complex.[48] The reason for this suggestion is not absolutely clear, but it appears to be based on the fact that on ultrasonication the lactose synthetase activity of the particles is lost, and can be reconstituted only by adding back high concentrations of the B protein.

Electron micrographs of the particles show clearly that they have a vesicular nature.[49] The well-known effect of sonication on vesicular structures is to disrupt the membranes, releasing the vesicular contents into the medium. The author is not aware of any instance where sonication has been used to disrupt the association of two proteins in a complex and he has suggested (Brew[50]) that the properties of the lactose synthetase particles are more simply explained by assuming that the N-acetyllactosamine synthetase component is attached to the inner surface of the vesicles, and that a solution of B protein is contained within the vesicles. On sonication, the disruption of the membranes will lead to release of the B protein into the surrounding medium, and its dilution in the locale of the A protein (Fig. 9), with a concomitant drop in lactose synthetase activity (see Fig. 3).

There are several reasons why a scheme of this type is attractive. As N-acetyllactosamine synthetase is in most tissues involved in the biosynthesis of secreted glycoproteins, its attachment to the inner surface of the Golgi membranes will place it directly in the secretory pathway of the cell and enable it to act on glycoproteins passing through the lumen of the Golgi bodies on their way to the exterior of the cell.[50] Consistent with this is the recent finding that Golgi membranes isolated from bovine liver homogenates have a high specific activity of N-acetyllactosamine synthetase.[51]

Furthermore, the fact that α-lactalbumin is a specifically secreted milk protein adds a further dimension to this scheme. Although N-acetyllactosamine synthetase is found in milk, its concentration is very much lower than that of α-lactalbumin. α-Lactalbumin is present in considerable quantities in milks, reaching a concentration as high as 5·5 mg/ml

in guinea-pig milk.[21] Evidence suggests that α-lactalbumin is synthesized continuously by the lactating mammary gland on the ribosomes
of the rough endoplasmic reticulum.[52] The protein then passes into the
lumen of the endoplasmic reticulum and eventually to the Golgi region
where it will meet the A protein and specify the synthesis of lactose.
Lactose synthesis thus can occur with N-acetyllactosamine synthetase
situated in the Golgi region as it is in other tissues. After interacting

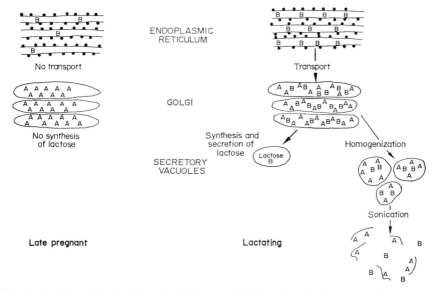

Fig. 9. A scheme showing the subcellular organization of the lactose synthetase system
in the mammary gland during pregnancy and lactation, and the implications of the
scheme for the control of lactose synthesis and the nature and properties of "lactose
synthetase particles" obtained from the lactating mammary gland. The lactose synthetase
A and B proteins are denoted by letters A and B.

with A protein, α-lactalbumin will be secreted from the cell along with
other milk proteins. This arrangement also provides a mechanism of
secretion for lactose, which, after synthesis in the Golgi region, can be
secreted in secretory vacuoles along with the milk proteins by a process
of reverse pinocytosis.[50]

   This scheme can be tested to see if it can account for the concentration
of lactose found in milk. If we assume: (1) that the contents of the Golgi
region are not diluted or concentrated on their way out of the mammary
cell in secretory vacuoles; (2) that the level of N-acetyllactosamine
synthetase is that found in the mouse mammary gland at the beginning

of lactation, i.e. 0·22 units (mμmol/min per mg of tissue)[41, 3] and that the Golgi region constitutes about 1% of the total tissue volume.

Calculation shows[53] that to account for the concentration of lactose in bovine (and other milks, see next section) each α-lactalbumin molecule will have to be in contact with NAL synthetase for about 5 min. The kinetics of secretion of α-lactalbumin by slices of lactating guinea-pig mammary gland show that it takes about 20 min for an α-lactalbumin molecule to be synthesized and secreted from the cell.[52] As a small protein like α-lactalbumin will be synthesized quite rapidly, a period of 5 min duration for its contact with the A protein is quite reasonable.

C. RELEVANCE OF ORGANIZATION TO CONTROL AND EVOLUTIONARY ORIGINS OF THE SYSTEM

Although it may appear wasteful to have an enzyme system of which one component is continuously secreted from the cell, when the sub-cellular organization of lactose synthetase is taken into account there are considerable advantages. If at any time, the synthesis and flow of α-lactalbumin through the Golgi region is stopped, the remainder of the protein will be secreted from the mammary cell and lactose synthesis terminated.[50] A further feature is that the establishment of secretory activity within the intracellular membranes of the epithelial cell during lactogenesis could be an important point of control, as such movement will be essential for bringing α-lactalbumin into contact with N-acetyl-lactosamine synthetase (see Fig. 9). Finally, it is interesting to note that this system of organization and control could have arisen automatically when the mutational changes in an ancestor of α-lactalbumin gave rise to lactose synthetase specifier activity. As this precursor was probably a secreted protein, it would have had the potentiality for interacting with an N-acetyllactosamine synthetase present on its secretory path-way, and previously active in glycoprotein synthesis alone.

D. INTERSPECIES VARIATION IN RATES OF LACTOSE SYNTHESIS

There is considerable variation in the relative concentrations of fat, lactose and protein in milk from different species. Two extremes for lactose are typified by some marsupials and aquatic mammals which secrete milk devoid of lactose, and man with a milk lactose content of 7%. In general there appears to be an inverse correlation between the milk lactose and protein content. Palmiter[54] has put forward the view that as α-lactalbumin is a milk protein, it will represent a fairly constant proportion of the total milk protein in different species and the milk α-lactalbumin concentration would appear to be related inversely to the lactose concentration. From this argument, it would appear that α-lactalbumin is unlikely to be concerned in the control of milk lactose synthesis.

Khatra and Brew[33] have used a different approach to this problem by examining the species variation in α-lactalbumins and N-acetyl-lactosamine synthetase. In this study α-lactalbumins have been isolated

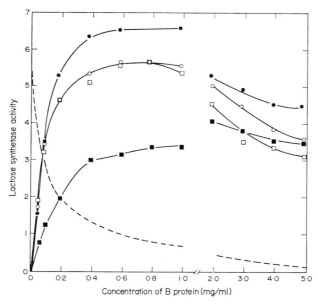

FIG. 10. The effect of increasing concentrations of four different B proteins on the lactose synthetase activity of human milk N-acetyllactosamine synthetase (Khatra and Brew[53]). ○, Human B protein; ●, bovine B protein; □, pig B protein; ■, guinea-pig B protein. The broken line shows the inhibition of N-acetyllactosamine synthetase activity with increasing concentration of human B protein.

from 4 species which show a considerable range in milk lactose content: human (7·0%), pig (4·5%), bovine (4·5%) and guinea-pig (2·7%), together with A proteins from two sources: human and bovine. Figure 10 shows the variation of lactose synthetase activity at increasing concentrations of these four α-lactalbumins. The resulting order of activities is related to the levels of lactose in the milks from which the B proteins were obtained. This order of activities for the B proteins is exactly the same when bovine instead of human A protein is used, and therefore does not represent mutual adaptation between the A and B proteins with respect to their interaction. The maximum velocities for lactose synthesis with the four different B proteins are directly proportional to the concentration of lactose in the milk from which the B protein was obtained.[53] The α-lactalbumin content in the milks, of these species at least, are at a level sufficient to saturate the A protein for lactose synthesis. It appears then, that if an exactly similar level and organization of N-acetyllactosamine synthetase is assumed in the mammary glands for these animals, a large part of the variation in the lactose content of the milks can be accounted for by variations in the structure and activity of the B proteins. As α-lactalbumin is synthesized along with the other milk proteins, the interspecies variation in its structure causes it to act as a modulator between the systems for synthesis of protein and lactose in the mammary gland, regulating the protein/lactose ratio and nutritional properties of different milks. Undoubtedly other factors will mediate in the regulation of lactose synthesis in the mammary gland, including the movement of substrates to the site of lactose synthesis. The activity variation in α-lactalbumin may be regarded as acting as a coarse control in the system, the other factors being concerned with the finer aspects of regulation.

IV. Discussion

In some ways, it is possible to regard lactose synthetase as a subunit enzyme arrested in an early stage of evolutionary development. Thus, it is composed of two proteins of separate evolutionary origin, one from a glycoprotein-synthesizing system, and the other probably derived from modification of a bacteriolytic enzyme. These proteins have not, however, progressed to the stage of mutual adaptation that is found in other similar systems, and may only interact through the mediation of a substrate molecule. Furthermore, the A protein, N-acetyllactosamine synthetase, persists in a dual biological activity, of which the glycoprotein synthesis aspect must be just as important as lactose synthesis. The structural adaptations associated with the new activity, lactose synthetase, appear to be all on the side of the specifier protein, α-lactalbumin. Not only is this protein of key importance in the evolutionary development of lactose synthesis, but it also plays an important role in controlling lactose synthesis during pregnancy and lactation, in linking together the biosynthesis of milk proteins and carbohydrate, and in controlling the organ-specific nature of lactose synthesis.

Many areas of this enzyme system obviously require further investigation. In particular we must await with great interest the outcome of X-ray crystallographic studies of α-lactalbumin and hope that the actual mechanism of action of α-lactalbumin and the relationship of its structure to activity will soon be elucidated.

In the author's laboratory, the structure and activity of α-lactalbumins from different species are being studied with a view to tracing the evolutionary pathway between α-lactalbumin and lysozyme and to obtain information about the effect of structural changes on the activity of the protein. Present results show that very large differences exist between the amino acid sequences of α-lactalbumins from different mammalian species (more than 25% of the residues being changed between pairs of proteins), which is consistent with the idea that the protein has a recent evolutionary origin and a high rate of evolutionary change.

We may hope that eventually further examples of genetically related groups of proteins with completely different biological functions will be found. However, the relatively recent divergence of the α-lactalbumin–lysozyme group leads us to hope that we may eventually obtain details of the amino acid substitutions which led to their functional separation.

REFERENCES

1. Long, C. A. (1969). The origin and evolution of mammary glands. *Bioscience,* **19,** 519–523.
2. Sloan, R. E., Jenness, R., Kenyon, A. L. & Regehr, E. A. (1961). Comparative biochemical studies of milks—I. Electrophoretic analysis of milk proteins. *Comp. Biochem. Physiol.* **4,** 47–62.
3. Dixon, G. H. (1966). Mechanisms of protein evolution. In *Essays in Biochemistry,* Vol. 2, pp. 147–204. Ed. by Campbell, P. N. & Greville, G. D. London: Academic Press.
4. Hansen, R. G., Wood, H. G., Peeters, G. J., Jacobson, B. & Wilken, J. (1962). Lactose synthesis VI. Labelling of lactose precursors by glycerol-1,3-C^{14} and glucose-2-C^{14}. *J. biol. Chem.* **237,** 1034–1039.
5. Kalckar, H. M. (1958). Uridine diphospho galactose: metabolism, enzymology and biology. *Adv. Enzymol.* **20,** 111–134.
6. Basu, S., Kaufman, B. & Roseman, S. (1965). Conversion of Tay-Sachs ganglioside to monosialoganglioside by brain uridine diphosphate D-galactose: glycolipid galactosyl transferase. *J. biol. Chem.* **240,** PC 4115–4117.
7. Gottschalk, A. (1969). Biosynthesis of glycoproteins and its relationship to heterogeneity. *Nature, Lond.* **222,** 452–454.
8. Morgan, W. T. J. & Watkins, W. M. (1969). Genetic and biochemical aspects of human blood group A-, B-, H-, Le^a- and Le^b-specificity. *Br. med. Bull.* **25,** 30–34.
9. Watkins, W. M. & Hassid, W. Z. (1962). The synthesis of lactose by particulate enzyme preparations from guinea-pig and bovine mammary glands. *J. biol. Chem.* **237,** 1432–1440.
10. Babad, H. & Hassid, W. Z. (1966). Soluble uridine diphosphate D-galactose: D-glucose β-4-D-galactosyl transferase from bovine milk. *J. biol. Chem.* **241,** 2672–2678.

11. Brodbeck, U. & Ebner, K. E. (1966). Resolution of a soluble lactose synthetase into two protein components and solubilization of microsomal lactose synthetase. *J. biol. Chem.* **241**, 762–764.
12. Brodbeck, U., Denton, W. L., Tanahashi, N. & Ebner, K. E. (1967). The isolation and identification of the B protein of lactose synthetase as α-lactalbumin. *J. biol. Chem.* **242**, 1391–1397.
13. Brew, K., Vanaman, T. C. & Hill, R. L. (1968). The role of α-lactalbumin and the A protein in lactose synthetase; a unique mechanism for the control of a biological reaction. *Proc. natn. Acad. Sci. U.S.A.* **59**, 491–497.
14. Crawford, I. P. & Yanofsky, C. (1958). On the separation of tryptophan synthetase into two protein components. *Proc. natn. Acad. Sci. U.S.A.* **44**, 1161–1170.
15. Brown, N. C. & Reichard, P. (1969). Role of effector binding in the allosteric control of ribonucleoside diphosphate reductase. *J. molec. Biol.* **46**, 39–55.
16. McGuire, E. J., Jourdian, G. W., Carlson, D. M. & Roseman, S. (1965). Incorporation of D-galactose into glycoproteins. *J. biol. Chem.* **240**, PC 4112–4115.
17. Kent, P. W. (1967). Structure and function of glycoproteins. In *Essays in Biochemistry*, Vol. 3, pp. 105–151. Ed. by Campbell, P. N. & Greville, G. D. London: Academic Press.
18. Hill, R. L., Brew, K., Vanaman, T. C., Trayer, J. P. & Mattock, P. (1968). The structure, function and evolution of α-lactalbumin. *Brookhaven Sympos. Biol.* **21**, pp. 139–154.
19. Sørensen, M. & Sørensen, S. P. L. (1939). The proteins in whey. *C.r. Trav. Lab. Carlsberg, Ser. Chim.* **23**, 55–99.
20. McKenzie, H. A. (1967). Milk Proteins. *Adv. Prot. Chem.* **22**, 55–234.
21. Brew, K. & Campbell, P. N. (1967). The characterization of the whey proteins of guinea-pig milk. The isolation and properties of α-lactalbumin. *Biochem. J.* **102**, 258–264.
22. Brew, K., Vanaman, T. C. & Hill, R. L. (1967). Comparison of the amino acid sequence of bovine α-lactalbumin and hen's egg white lysozyme. *J. biol. Chem.* **242**, 3747–3749.
23. Jukes, T. H. & Cantor, C. R. (1969). Evolution of Protein Molecules. In *Mammalian Protein Metabolism*, Vol. 3, pp. 21–132. Ed. by Munro, H. N. New York and London: Academic Press.
24. (a) Canfield, R. E. & Liu, A. K. (1965). The disulfide bonds of egg white lysozyme (Muramidase). *J. biol. Chem.* **240**, 1997–2002.
24. (b) Jollès, P. (1967). Relationship between chemical structure and biological activity of hen egg-white lysozyme and lysozymes of different species. *Proc. R. Soc. B*, **167**, 350–364.
25. Howard, J. B. & Glazer, A. N. (1969). Papaya lysozyme. Terminal sequences and enzymatic properties. *J. biol. Chem.* **244**, 1399–1409.
26. Inouye, M. & Tsugita, A. (1966). The amino acid sequence of T4 bacteriophage lysozyme. *J. molec. Biol.* **22**, 193–196.
27. (a) Canfield, R. E. (1970). Personal communication.
27. (b) Jollès, T. & Jollès, P. (1969). La structure chimique du lysozyme du lait de femme: etablissement d'une formule developpée provisiore. *Helv. chim. Acta*, **52**, 2671–2675.
28. Shotton, D. M. & Watson, H. C. (1970). Three dimensional structure of tosylelastase. *Nature, Lond.* **225**, 811–816.

29. Perutz, M. F., Muirhead, H., Cox, J. M. & Goaman, L. C. G. (1968). Three-dimensional Fourier synthesis of horse oxyhaemoglobin at 2·8 Å resolution: the atomic model. *Nature, Lond.* **219**, 131–139.

30. Blake, C. C. F., Mair, G. A., North, A. C. T., Phillips, D. C. & Sarma, V. R. (1967). On the conformation of the hen egg-white lysozyme molecule. *Proc. R. Soc. B*, **167**, 365–377.

31. Browne, W. J., North, A. C. T., Phillips, D. C., Brew, K., Vanaman, T. C. & Hill, R. L. (1969). A possible three-dimensional structure of bovine α-lactalbumin based on that of hen's egg-white lysozyme. *J. molec. Biol.* **42**, 65–86.

32. Aune, K. (1968). Ph.D. Dissertation, Duke University.

33. Kronman, M. J. (1968). Similarity in backbone conformation of egg white lysozyme and bovine α-lactalbumin. *Biochem. biophys. Res. Commun.* **33**, 535–541.

34. Cowburn, D. A., Bradbury, E. M., Crane-Robinson, C. & Grazer, W. B. (1970). An investigation of the conformation of bovine α-lactalbumin by proton magnetic resonance and optical spectroscopy. *Eur. J. Biochem.* in press.

35. Atassi, M. Z., Habbeb, A. F. S. A. & Rydstedt, L. (1970). Lack of immunochemical cross-reaction between lysozyme and α-lactalbumin and comparison of their conformations. *Biochim. biophys. Acta*, **200**, 184–187.

36. Tanahashi, N. Brodbeck, U. & Ebner, K. E. (1968). Enzymic and immunological activity of various B proteins of lactose synthetase. *Biochim. biophys. Acta*, **154**, 247–249.

37. Arnon, R. & Sela, M. (1969). Antibodies to a unique region in lysozyme provoked by a synthetic antigen conjugate. *Proc. natn. Acad. Sci. U.S.A.* **62**, 163–170.

38. Blake, C. C. F., Johnson, L. N., Mair, G. A., North, A. C. T., Phillips, D. C. & Sarma, V. R. (1967). Crystallographic studies of the activity of hen egg-white lysozyme. *Proc. R. Soc. B*, **167**, 378–388.

39. Castellino, F. J. & Hill, R. L. (1970). The carboxymethylation of bovine α-lactalbumin. *J. biol. Chem.* **245**, 417–424.

40. Findlay, J. B. C. & Brew, K. Unpublished results.

41. Turkington, R. W., Brew, K., Vanaman, T. C. & Hill, R. L. (1968). The hormonal control of lactose synthetase in the developing mouse mammary gland. *J. biol. Chem.* **243**, 3382–3387.

42. Palmiter, R. D. (1969). Hormonal induction and regulation of lactose synthetase in mouse mammary gland. *Biochem. J.* **113**, 409–417.

43. Kuhn, N. J. (1968). Lactogenesis in the rat. Metabolism of uridine diphosphate galactose by mammary gland. *Biochem. J.* **106**, 743–748.

44. Juergens, W. G., Stockdale, F. E., Topper, Y. J. & Elias, J. J. (1965). Hormone-dependent differentiation of mammary gland in vitro. *Proc. natn. Acad. Sci. U.S.A.* **54**, 629–634.

45. Turkington, R. W., Lockwood, D. H. & Topper, Y. J. (1967). The induction of milk protein synthesis in post-mitotic mammary epithelial cells exposed to prolactin. *Biochem. biophys. Acta*, **148**, 475–480.

46. Turkington, R. W. & Hill, R. L. (1969). Lactose synthetase: progesterone inhibition of the induction of α-lactalbumin. *Science, N.Y.* **163**, 1458–1460.

47. Coffey, R. G. & Reithel, F. J. (1968). The lactose synthetase particles of lactating bovine mammary gland. Preparation of particles with intact lactose synthetase. *Biochem. J.* **109**, 169–176.

48. Palmiter, R. D. (1969). Properties of lactose synthetase from mouse mammary gland: role of a proposed third component. *Biochim. biophys. Acta,* **178,** 35–46.

49. Coffey, R. G. & Reithel, F. J. (1968). The lactose synthetase particles of lactating bovine mammary gland. Characteristics of the particles. *Biochem. J.* **109,** 177–183.

50. Brew, K. (1969). Secretion of α-lactalbumin into milk and its relevance to the organization and control of lactose synthetase. *Nature, Lond.* **223,** 671–672.

51. Fleischer, B., Fleischer, S. & Ozawa, H. (1969). Isolation and characterization of Golgi membranes from bovine liver. *J. cell. Biol.* **43,** 59–79.

52. Brew, K. & Campbell, P. N. (1967). Studies on the biosynthesis of protein by lactating guinea-pig mammary gland. Characteristics of the synthesis of α-lactalbumin and total protein by slices and cell-free systems. *Biochem. J.* **102,** 265–274.

53. Khatra, B. S. & Brew, K. (1970). Manuscript in preparation.

54. Palmiter, R. D. (1969). What regulates the lactose content in milk? *Nature, Lond.* **221,** 912–914.

Metabolite Transport in Mitochondria:
An Example for Intracellular Membrane Function

M. KLINGENBERG

*Institut für Physiologische Chemie und Physikalische Biochemie,
Universität München, 8000 München 15, Germany*

I. Introduction

This essay is intended to give a survey of the relatively new field
of metabolite transport in mitochondria. Research in this highly active
field is still at an early stage; our knowledge is therefore limited, and
development is rapid and subject to change. The present account is
not meant to provide a comprehensive review of this vast field but rather

Abbreviations. AdN, adenine nucleotide; ASPM, N-(N-acetyl-4-sulphamoylphenyl)-
maleimide; DTNB, 5,5′-dithio-bis(2-nitrobenzoic acid); FCCP, (p-trifluoromethoxy)-
carbonylcyanide-phenylhydrazone; NEM, N-ethyl-maleimide; ADP-N-P and ADP-C-P,
ADP-imidophosphate and ADP-methylene phosphate.

to focus on fundamental developments and viewpoints. Thus the principles for understanding metabolite transport, the methodological approaches and some possible models for its mechanism are stressed. A brief review of this subject has been presented elsewhere.[1]

II. Principles of Closed Space

At the cellular level it is useful to speak of closed spaces analogous to closed systems in order to describe some of the principles of construction in metabolic sequences. The cellular structures which define the closed spaces are the membranes. Their permeability properties set a barrier both to the constituents of the closed space and to the majority of the metabolites. Inside the closed space the metabolism is determined by a complement of enzymes and coenzymes. The chemical substrates of the catalytic machinery are also largely confined within the closed space. The metabolites are maintained within the closed space at a concentration which approximates to that of a steady state. Since the steady state can be maintained and supported only by a certain uptake and release of metabolites, the closed system must be equipped with certain entrances and exits.

The function of the membrane is therefore a dual one, on the one hand serving as a general barrier and, on the other, allowing certain permeabilities or "pores". Both functions must be accounted for by the structure of the membranes. The role of providing a general impermeable wall is a relatively crude one and, in principle, is achieved by the lipid bi-layer. The function in permitting the specific uptake and release of substrates is much more subtle and is achieved by the special properties of biological membranes which differentiate them from those artificially prepared.

As will be shown later, the selective permeability is the result partially of an unspecific "porosity" of the membrane, particularly for low molecular weight uncharged compounds, coupled with the effect of specific carriers. The main interest of the discussion will be focused on the carrier-catalysed transport.

The mitochondrion is a particularly interesting example of a selectively closed system, being enclosed by two membranes, only one of which, the inner membrane, retains sufficient impermeability to form a closed system, at least in isolated preparations of mitochondria. The enclosed space which contains the impermeable constituents is called the matrix space. Up to this point, the inner membrane has been defined merely as a barrier between two spaces. It is useful, for a more detailed analysis of the membrane function and in particular the mechanism of transport, to differentiate between various phases, such as the matrix phase, inner membrane phase and outer phase, which in the mitochondria is often called the intermembrane space, being located between the inner and outer membranes (Fig. 1). A considerable portion of mitochondrial reactions take place in the membrane phase. It may be postulated that the localization in the membrane phase of any reaction is linked to a vectorial function, such as coupled transport of a metabolite and of ions. Only in this circumstance is the major property of the membrane —the separation of spaces—effectively exploited by the membrane-fixed catalysts.

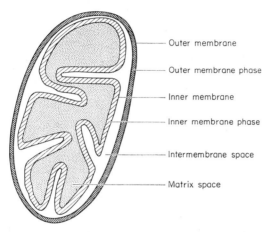

FIG. 1. The mitochondrion.

The major metabolic systems in mitochondria are concerned with the oxidative degradation of substrate, linked to oxidative phosphorylation. The substrate degradation reactions are mainly localized in the matrix, whereas the hydrogen transfer to oxygen comprising the respiratory chain, a great number of dehydrogenases and the system of oxidative phosphorylation are membrane-bound. The relatively high proportion of membrane-bound reactions is reflected in the mitochondria in an unusually high ratio of membrane surface per volume, approximately 50 m^2/ml in mitochondria from heart or insect flight muscle. This is an important prerequisite for the favourable conditions which permit studies of membrane function in the mitochondria.

For the present purpose three types of membrane reaction may be differentiated: 1. transport through the membrane without chemical modification (substrate translocation); 2. transport of groups through

5*

the membrane coupled to splitting or formation of chemical bonds (group translocation); 3. transport of cations and electrons through the membranes with or without coupling to electron transport. The following discussion will concentrate on type 1, the transport of metabolites through the inner mitochondrial membrane, with some excursions into type 2, group translocation.

III. Experimental Approach to Studying Permeability and Transport in Mitochondrial Membranes

In the mitochondria, the experimentalist faces the problem of studying penetration from a relatively large volume (the suspension medium) into a volume which is orders of magnitude smaller. Therefore the intramitochondrial amount of substance, of which the uptake is to be determined, will be expected to be rather small if equilibration occurs. A highly permeable substance, the penetration of which is mainly diffusion-limited, will equilibrate with the intramitochondrial space in less than a millisecond. To this category belong H_2O, O_2, CO_2, ethanol, NH_3, acetate, etc. In general, the rate of permeation of "highly permeable" substances into mitochondria has not been measured. Substances of "limited permeability" penetrate relatively slowly, due to a penetration barrier in the membrane. Limited permeability often involves transport by a specific carrier, and in such cases it may be possible to measure the rates of penetration.

A. DIRECT MEASUREMENT OF TRANSPORT

A "direct" method for measuring transport and permeability consists of localizing the substances in the intra- and extramitochondrial spaces (cf. Table 1). Either the kinetics of transport or the steady state distribution between the two spaces can be measured. For these measurements the mitochondria have to be separated from the suspension medium, in general by sedimentation or by filtration. It may be required to preserve the metabolic state by combining the separation with a simultaneous quenching of the mitochondria or supernatant. Often these requirements are not fulfilled, and only the distribution of radioactive label is measured. In this case quantitative information on the concentration gradient can be obtained only if the permeant substance is not metabolized.

In order to preserve the metabolic state during separation, the sedimentation is combined with filtration through a non-aqueous layer of silicone into an acid or alkaline layer. The resulting immediate quenching prevents alterations of the metabolic state of the penetrated substance

TABLE 1

Methods of measuring transport in mitochondria

Methods	Quench possibility	Time resolution	Application and remarks
I. *Direct assay of transported substances (separation methods)*			
Cellulose pore filtration	External space	1 s	Most rapid separation. Quenching of external compartment possible when injected into acid. Problem of leakage from mitochondria accumulated on the filter[2]
Centrifugal filtration			
Sucrose density layer	None		The equilibration is disturbed because external permeant is removed[3]
Silicone layer	Internal space	40 s	Preservation of instantaneous internal metabolite level. Swinging bucket rotors of ultracentrifuge or microcentrifuge with 90° angle[2, 4–7]
Multiple type layer	Internal space	5–20 s	Transport is started during centrifugation by bringing mitochondria from storage to incubation layer. Then separation through silicone layer[2, 6, 8]
Centrifugal sedimentation	None	20 s	Simple method without quenching. Careful correction of adherent external space required. Fastest in microcentrifuge
Inhibitor stop of transport (centrifugal sedimentation)			
Programmed rapid sampling	None	0·3 s	Transport stopped by inhibition. Requires that $t_{1/2}$ of inhibition reaction is at least 10 times quicker than $t_{1/2}$ of transport[2,9,10]
Stopped flow		20 ms	Inhibitor addition by mixing chambers. No quenching, careful control for uninhibited transport.
II. *Indirect methods*			
Osmotic swelling		1 s	For uptake of ions the addition of permeable counter-ion, e.g. NH_4^+,[11] K^+ + nigericin, or valinomycin[12]
Interaction with intramitochondrial enzymes			Rate-limiting step often questionable

and its products, which would be expected to occur rapidly under the altered (anaerobic) conditions in simple centrifugal sedimentation or in cellulose pore filtrations. In many cases, however, this may not be important if the alterations are negligible and tracer amounts of radioactive substances are used.

The combined quenching and separation is required also for measuring the specificity of transport from the inside, as will be discussed for adenine nucleotide translocation. In this and other cases it may be important to quench the extramitochondrial fluid simultaneously with the separation, in order to assay the released or permeant metabolites in the extramitochondrial fluid, which contains released enzymes. The cellulose pore method of filtration permits instantaneous quenching by rapid passage of the emerging filtrate into acid.

The method of centrifugal filtration and sedimentation is usually too slow for following the kinetics of permeation. Permeation times as short as 20 s can be obtained by introducing an intermediate layer, where the mitochondria are exposed to the permeant substrate only during centrifugation when they pass through this layer.[6] The depth of this layer determines the time of permeation. Considerably shorter times can be obtained by rapid removal of the mitochondria from the suspension by pressure filtration through membrane filters. With a special apparatus for rapid addition, and initiation of pressure, penetration time could be reduced to 2 s.[2]

Permeation times down to 0·2 s can be obtained by means of inhibitors of the translocation. In this case the reaction is started by rapid mixing of the permeant substance with the mitochondria and is then stopped by addition of the transport inhibitor in a special device. The mitochondria are subsequently sedimented and assayed for the permeant substance. This method is limited to those cases where an effective inhibitor of transport of a particular substance is available, such as atractyloside for the adenine nucleotide translocation. With rapid and consecutive sampling and immediate mixing of the samples with the inhibitor, the time sequence of the permeation can be determined.

B. INDIRECT METHODS

The indirect methods make use either of the osmotic effects of permeation or of the interaction of permeated substances with components localized in the matrix or the inner side of the membrane (cf. Refs. 11 and 12). These methods have been quite convenient for elucidating the permeability of a number of substances, since recording techniques can be applied to follow both the osmotic effects and the kinetics of the enzymic reactions in the mitochondria. The osmotic effects require high concentrations of permeant substrates, so that often isotonic concentrations are used which are orders of magnitude above their physiological level. The massive uptake of permeant substances by the mitochondria is accompanied by influx of water and swelling, so that the resulting decrease in light scattering by the mitochondria can be recorded conveniently. If the permeant substance is an anion, conditions must be devised such that the permeation of the substance under study, and not that of a concomitant cation, is limiting. This condition can be fulfilled by supplying the mitochondria with a high concentration of NH_4^+, e.g. the NH_4^+ salt of the permeant anion, since NH_4^+ is freely permeable due to its dissociation to NH_3. With this method useful data on the anion permeability have been obtained.[11-13] The osmotic method measures only a net uptake, and therefore does not permit one to follow exchange transport.

The lack of reactivity of substrates with mitochondrial enzymes was the first indication of a specific permeability barrier. However, it had to be established that the reactive enzymes are truly intramitochondrial and are not located on the surface of the cristae membrane, as was recently shown for some dehydrogenases. In many cases redox reactions of the intramitochondrial NAD or NADP system can be used as convenient indicators of permeation, since their absorption or fluorescence changes can be conveniently recorded (cf. Ref. 14). In contrast to the osmotic technique, low (i.e. physiological) concentrations of the substances can be used for the permeation assay. However, the rate of the redox reactions may not reflect correctly the rate of permeation but rather be limited by the dehydrogenase activity.

IV. Metabolite Transport Systems

A. GENERAL

The study of metabolite transport through the mitochondrial membrane was started relatively late compared with studies on transport in bacteria, erythrocytes or other single cell systems. The direction of research on mitochondria has been determined by their enormously high metabolic activity, in particular in oxidative metabolism. Therefore, in general, transport was approached rather from the viewpoint of its relation to intramitochondrial reactions than as an investigation of transport *per se*. Transport reactions in mitochondria are often one to two orders of magnitude faster than those in whole cells and the determination of transport rates is correspondingly more difficult. Consequently, in many cases only qualitative results have been obtained.

5**

Mitochondrial transport is, however, at an advantage with respect to the elucidation of carrier systems. The high rate of metabolite transport indicates that these carriers are present in the membrane at high concentrations, which should greatly facilitate their identification and eventual molecular characterization. Indeed, some carriers have already been defined and their molecular identification has been approached.

A schematic survey of the metabolite fluxes through the mitochondrial membrane is depicted in Fig. 2. In principle, in the same manner as in an intact cell, a selected number of substances are taken up or released from the mitochondria which, with their largely impermeable membrane, retain not only enzymes and coenzymes but also cations and certain metabolites.

Mitochondrial metabolite transport can be classified from the functional point of view into a more fundamental system which effects the transport of the same metabolites for all mitochondria irrespective of

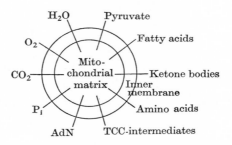

FIG. 2. The metabolite flux through the mitochondrial membrane (TCC = tricarboxylic acid cycle).

source, and particular systems related to the specialized function of the individual cell. The fundamental transport system includes the permeation of O_2, H_2O, CO_2, phosphate, adenine nucleotides and essential substrates such as pyruvate and fatty acids, whereas the transport of tricarboxylic acid cycle intermediates can be considered as being linked to the more special tasks of the mitochondria. Some of these metabolites permeate unspecifically through the membrane and do not require a carrier, whereas others can penetrate the membrane only by means of specific carriers provided for them in the mitochondria. The mere existence of these specific transport systems, which have been detected *in vitro*, implies that they are probably also important *in vivo*, i.e. in the intact cell (see also Refs. 13, 13a, 13b, 14a, 14b). The main theme in the following discussions will concern these specific carrier systems.

B. UNSPECIFIC OR NON-CARRIER-LINKED METABOLITE TRANSPORT

Oxygen has to penetrate at least an essential part of the mitochondrial membrane in order to reach its reaction site at the cytochrome oxidase. This can be inferred from the conclusion that the binding site of azide to the cytochrome oxidase is inside the generally impermeable barrier.[15] The increased solubility of oxygen in the lipids may cause oxygen to diffuse along the cristae membrane from the periphery of the mitochondrion into the inner part along the cristae invaginations and thus reach the cytochrome oxidase closely linked to the inner site of the lipid phases.

Water is generated at the rate of about 300–1000 μmol/min per g protein corresponding to a maximum efflux of about 5% of the intramitochondrial H_2O per min. Much higher rates of water fluxes in mitochondrial membranes are observed during osmotic swelling. Thus the intramitochondrial H_2O is increased by 200 to 300% in a few seconds on the transition from an isotonic to a hypertonic medium. It may be concluded that the efflux of water as a result of the oxidation of the substrate by oxygen is not limited by a diffusion barrier.

Carbon dioxide transport from the inside to the outside of mitochondria is not yet fully explored. The occurrence of a carbonic anhydrase (EC 4.2.1.1) in mitochondria has been reported[16] and an efflux of CO_2 as HCO_3^- was suggested, possibly even in exchange for P_i. However, in swelling experiments with high concentrations of NH_4^+ and HCO_3^- no uptake of HCO_3^- was noted.[17] A function of carbonic anhydrase in providing HCO_3^- for permeation has also been discussed in later work.[18] Actually, no significant barrier for the penetration of CO_2 through the mitochondrial membrane would be expected. The mitochondrial membrane appears to permit a rapid carrier-free diffusion of a number of metabolically important monocarboxylic acids, such as pyruvate, acetoacetate, hydroxybutyrate, short and long chain fatty acids. No quantitative data on the permeability of the monocarboxylates are available, apart from some relative estimations from swelling, to the effect that acetate and propionate penetrate considerably faster than pyruvate, hydroxybutyrate and acetoacetate.

The access of *pyruvate* to its dehydrogenase in mitochondria is of great interest in view of the central role of pyruvate oxidation in carbohydrate catabolism. No evidence for the occurrence of a transport carrier for pyruvate, according to the criteria discussed below, has been produced. It is therefore feasible that the observed uptake is a carrier-free unspecific process similar to that for the other monocarboxylates.

The α-keto group decreases the permeation rate as shown by comparison with propionate. It might be speculated that the pyruvate dehydrogenase complex faces the outer surface of the inner membrane, as does the glycerolphosphate dehydrogenase.[19] However, in contrast to glycerolphosphate dehydrogenase there is no need for an outside localization of the pyruvate dehydrogenase in view of the permeability of pyruvate. Similar reasoning applies to the penetration of ketone bodies, hydroxybutyrate and acetoacetate, the dehydrogenases of which are also tightly bound to the membrane. At any rate, these substrates interact with the intramitochondrial nicotinamide nucleotide system and are used as indicators of its redox potential[20] (cf. also Refs. 21 and 22).

It appears appropriate at this stage to discuss the reason why the membrane is unspecifically permeable to monocarboxylates but impermeable to dicarboxylates and tricarboxylates unless a carrier is

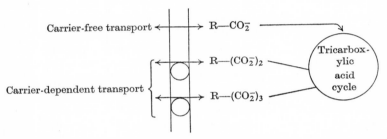

FIG. 3. Mono-, di- or tricarboxylates as permeant "feeders" or impermeant members of the tricarboxylic acid cycle.

provided. In the simplest approximation the permeability can be considered to be proportional to the concentration of the permeant in the membrane. Only the undissociated acids are significantly dissolved in the lipid phase; therefore, the concentration in the membrane is approximately proportional to that of the undissociated acid in the aqueous phase. In the monocarboxylates the proportion of undissociated to total acid is $HA/A_o = 10^{-4}$ to 10^{-3}, and for dicarboxylates $H_2A/A_o = 10^{-6}$ to 10^{-5}, at pH 7. Between the homologues acetate and succinate the permeability should therefore decrease by 10^{-3}. This ratio may be even lower due to entropy effects caused by the hydration difference between mono- and dicarboxylates which would decrease the distribution coefficient (cf. discussion in Ref. 23). The net result of these factors is that the dicarboxylates and tricarboxylates are "practically" impermeable, in contrast to the monocarboxylates.

In view of these conclusions, it is interesting to note that the main entering substrates, such as pyruvate and fatty acids, as well as the extruded products, such as ketone bodies, are monocarboxylates, and the intermediates of the tricarboxylic acid cycle consist of non-penetrant dicarboxylates and tricarboxylates (cf. Fig. 3). Thus the intermediates of the tricarboxylic acid cycle are confined to the matrix space where the corresponding enzymes are located, an advantage for an effective functioning of the cycle. It is intriguing to speculate that for this very reason the terminal oxidation cycle has evolved so as to utilize substrates which consist only of di- or tricarboxylates. In contrast, the feeders of the cycle are the permeable monocarboxylates. This assumption does not contradict the existence of carriers for the intermediates which are provided whenever an interaction of the intermediates with the cytoplasm has become necessary.

C. CARRIER-DEPENDENT METABOLITE TRANSPORT
(cf. Table 2)

The uptake and release of substrates (P_i and ADP) and products (ATP) of oxidative phosphorylation are fundamental functions of mitochondria. The transport of the adenine nucleotides and of phosphate have separate carriers. The transport of phosphate will be treated here in connexion with that of the dicarboxylates and tricarboxylates with which it is closely linked. The transport of adenine nucleotides will be discussed in a separate section because of the large amount of information available on this system.

The transport of non-permeant metabolites is catalysed by specific carriers which are apparently present in the mitochondrial membrane in response to the specific requirements of the interaction between mitochondria and cytoplasm. A number of criteria have been applied to characterize the transport of these metabolites as a carrier-driven process. The most direct and simple criterion is the existence of a specific inhibitor. More complex criteria apply to the kinetics of transport, e.g. saturation characteristics of the concentration-dependence. The existence of high specificity is a major criterion for the participation of a carrier.

The major transport reactions in mitochondria which are identified according to these criteria as carrier-driven transport belong to oxidative phosphorylation and to the tricarboxylic acid cycle. Further carriers have been identified for amino acids. The uptake and release of substrates and products of oxidative phosphorylation are fundamental functions of mitochondria. Therefore the carriers linked to oxidative phosphorylation have been found in all mitochondria tested so far. In contrast, carriers of the tricarboxylic acid cycle intermediates and, in

TABLE 2

Survey of major metabolite transport systems in mitochondria

Metabolic context	Permeant	Carrier	Exchange	Inhibitor	Ref.	Remarks
Oxidative phosphorylation	ADP	+	Mutual	Atractyloside	24	"Group translocation", carnitine catalysed
	ATP	+	—	Bongkrekic acid	25, 26	
	P_i		—	SH reagents	27, 28	
Feeders to terminal oxidation	Pyruvate	—	—			
	Fatty acids	—	—			
Ketone bodies	Hydroxybutyrate	—	—			
	Acetoacetate	—	—			
Tricarboxylic acid cycle intermediates						
Dicarboxylates	Malate	+	With P_i (tricarboxylates)	Butylmalonate	29	Exchange with P_i, citrate, isocitrate
	Succinate			Iodobenzyl-malonate	30	
	Ketoglutarate	+	With dicarboxylates			Exchange with malate, succinate, malonate
Tricarboxylates	Citrate	+	With dicarboxylates	2-Ethylcitrate	31	Exchange with malate, succinate
	Isocitrate					
Amino acids	Aspartate	+	—			Activated by glutamate, ketoglutarate
	Glutamate	+	—	Avenaciolide	32	"Group translocation"

particular, of amino acids are found only in mitochondria from certain cells. The fact that the occurrence of some metabolite transport is confined to the mitochondria of certain cells is a further indication of the specific nature of a carrier-linked translocation.

1. *Transport of Phosphate*

The transport of P_i is treated separately here from the adenine nucleotide (AdN) translocation, although both substrates are involved in oxidative phosphorylation. However, the P_i carrier appears to be less closely linked to the AdN transport than to the transport of dicarboxylates and tricarboxylates, where it holds the key position.

The high activity of P_i transport into mitochondria was utilized early in studies on cation transport, where P_i was a convenient anion which compensated for the uptake of cations. Considerable information about the P_i transport was gained using the indirect method of following mitochondrial swelling. Judging from this method, P_i permeates considerably more slowly than acetate,[17] but at a rate about equal to that of the dicarboxylates. The permeation of P_i is strongly facilitated in the presence of NH_4^+ ions, indicating that the undissociated acid accompanies the diffusion of NH_3 across the membrane. K^+ plus P_i diffuse across the

$$NH_4^+ \rightleftharpoons H^+ + NH_3 \longrightarrow NH_3 \xrightarrow{\quad} NH_4{}^+$$
$$P_i^- + H^+ \rightleftharpoons P_i \longrightarrow P_i \xrightarrow{H^+} P_i^-$$

membrane more rapidly when both uncoupler (FCCP) and a K^+-ionophore (valinomycin) are present.[12] This is interpreted as an exchange of the H^+ taken up with P_i against external K^+:

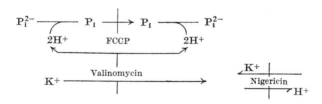

An interesting feature of the P_i transport is its inhibition by SH reagents.[27, 28] The involvement of SH groups indicates the carrier nature of the P_i transport, particularly since a great number of carrier-dependent transport mechanisms in other membranes have been defined as SH-dependent. In the mitochondria, among all known carriers, so far only the P_i carrier has turned out to be SH group dependent. This permits the elucidation of the influence of P_i on the transport of other metabolites, in particular its key position for the entry of dicarboxylates and tricarboxylates, as will be discussed below.

2. *Transport of Tricarboxylic Acid Cycle Intermediates*

The permeation of dicarboxylates has been noted indirectly through redox changes of the intramitochondrial nicotinamide nucleotides. In the case of succinate the rate of substrate oxidation could be quantitatively correlated to the rate of permeation and to the steady state of the internal to external distribution of succinate.[33] Direct studies by methods utilizing filtration centrifugation have given a more detailed picture of the distribution of these anions between the inner and outer membranes, and of their competition (cf. Refs. 5, 33, 34, 35 and 36).

The various carriers were first elucidated when the convenient but relatively crude method of osmotic swelling was applied to follow the kinetics of the substrate entry.[11, 17] In this way a number of criteria essential for the identification of carriers were probed, such as specificity, the existence of inhibitors and the peculiar interaction between the transport of various substrates. Three different carriers for the permeation of the tricarboxylic acid cycle intermediates have been discriminated (cf. Ref. 37 and Fig. 3): (a) for dicarboxylates (malate, succinate); (b) for ketoglutarate; (c) for tricarboxylates (isocitrate, citrate). Fumarate and oxaloacetate are found not to permeate easily by any carrier.[11] In the absence of competition by other dicarboxylates and with Ca^{2+}, oxaloacetate is said to be transported also by the dicarboxylate carrier.[38]

The dicarboxylic entry is most active with L-malate, less active with succinate and has some activity with malonate. The net uptake of dicarboxylates requires the presence of P_i. Among the tricarboxylates, isocitrate, citrate and *cis*-aconitate require malate for activation before being taken up. For the transport of ketoglutarate a carrier apparently different from those for dicarboxylates and tricarboxylates is defined. Ketoglutarate entry is not activated by P_i, but requires malate, malonate or succinate. Tricarboxylic acid transport is also activated by malate and—in contrast to ketoglutarate transport—by isomalate though not by malonate.

The activation of the dicarboxylate and tricarboxylate transport by P_i, malate, etc. can be explained on the basis of an exchange (see Fig. 4). Thus malate can enter only in exchange for P_i. Ketoglutarate or citrate is taken up only in exchange against malate, etc. (see Table 2). A 1:1 exchange between added and intramitochondrial metabolites has been directly demonstrated[39, 40, 41] using centrifugal filtration and sedimentation techniques, following the procedures established for the AdN exchange.[2] An equivalent release of malonate in exchange for P_i or ketoglutarate has been found. Furthermore, the exchange between other combinations such as malate–citrate, P_i–malate, etc. has been demonstrated in connexion with studies on the effect of inhibitors and parallel studies on cation and H^+ movements.[42]

It appears, therefore, that for the metabolite transport through the mitochondrial membrane the principle of exchange is realized not only for AdN but also for other important metabolites. It is clear that by an obligatory exchange, a state of osmotic equilibrium with respect to metabolites is attained. Even the uncompensated uptake of P_i does not disturb the osmotic equilibrium of metabolizing mitochondria, since P_i is incorporated into ATP.

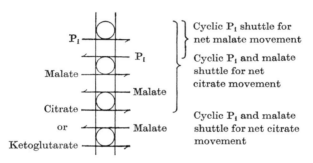

FIG. 4. The interlocked transport of P_i, di- and tricarboxylates.

As a result of the exchange, the four metabolite transport systems interact in a cascade-like manner (Fig. 4). P_i is the key substrate since it can be transported without exchange. The net uptake of dicarboxylates such as malate is possible when the malate–P_i exchange is coupled to the net P_i transport. A catalytic amount of P_i can serve as a cyclic shuttle leading to a net uptake of malate. A net uptake of ketoglutarate or citrate (tricarboxylates) requires interaction between the tricarboxylate–dicarboxylate exchange and the P_i shuttle. Thus a combined cyclic dicarboxylate–P_i shuttle can catalyse the net uptake of these substrates. The net uptake of di- and tricarboxylates observed in mitochondria without addition of P_i also appears to obey this mechanism, since P_i may originate from endogenous P_i of the mitochondria. In fact, the net uptake of substrates in mitochondria, such as the uptake of succinate, etc., is inhibited by SH reagents which primarily inhibit the P_i transport and therefore indirectly the net uptake of other substrates as well.

D. TRANSPORT OF AMINO ACIDS

Of the two major amino acids linked to mitochondrial metabolism only glutamate appears to be transported by a specific carrier.[43] Apparently its transport does not require an activating exchange metabolite. Of considerable interest is the existence of a specific inhibitor for the glutamate carrier, the antibiotic avenaciolide.[32] The inhibition appears to be competitive, in agreement with some structural similarity between avenaciolide and glutamate.

It is interesting that aspartate does not appear to be translocated as unmodified substrate,[44] but is converted during the transport to oxaloacetate. As an example for group translocation, it will be discussed in Section V.

E. CARRIER MODELS FOR METABOLITE TRANSPORT

The simplest interpretation of the interacting metabolite transport reactions is to assume an ambivalent specificity of the various carriers. According to this model each substrate would be transported by several

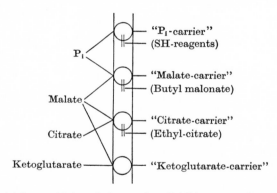

FIG. 5. The multiple specificity of the carriers (inhibitors are shown in parentheses).

carriers (Fig. 5): P_i both by the P_i and dicarboxylate carriers, malate by the three carriers for dicarboxylates, for citrate and for ketoglutarate.

According to a single site mobile carrier model, the dicarboxylate carrier, for example, is either charged with malate or with P_i (Fig. 6). For a compulsory exchange the carrier can make the translocation step only if either the site for P_i or that for malate is occupied. An unsatisfactory consequence of this model is that in principle the binding site should be specific for such different substrates as, for example, P_i and L-malate, although it is highly sensitive to variations in the structure of the dicarboxylates. It appears more feasible to assume that the two exchanging metabolites bind on different sites of the carrier. A further step would be the assumption that the two different binding sites are located on opposite sides of the carrier ("double carrier" model, Fig. 6). An exchange would be achieved by reorientating the carrier so that its two binding sites are now directed to opposite sides of the membrane. Only the carrier occupied by the two different substrates would be able to make the reorientation step in order to achieve compulsory exchange.

It might be speculated that the double carrier is composed of two half carriers, each specific either for P_i or malate. The half carrier for P_i might be identical both in the P_i and the malate–P_i carrier and the half carrier for malate the same as in other exchange carriers. A common pool of half carriers for malate or for P_i might combine with the half carrier of P_i or of citrate to form the twin carriers active in exchange.

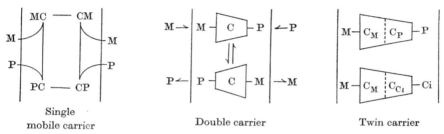

Single
mobile carrier

Double carrier

Twin carrier

FIG. 6. Carrier models for the interlocking metabolite exchange. M, Malate; P, phosphate; Ci, citrate; C, carrier. Arrows indicate direction of transfer.

In this model the variety of protein chains would be minimized and it would be accompanied by an economy in the information required for synthesis of the carriers.

From transport studies it is difficult to make a choice between these models. All these models can catalyse both a homo- (between identical substrates) and a hetero-exchange (between different substrates). A closer elucidation of these models must await studies on the level of the carriers, identification of the binding sites, etc. as has been successfully commenced for the AdN carrier, to be discussed below. The difference in the specificity of dicarboxylates in the citrate–malate exchange[14] (isomalate) and in the malate–ketoglutarate exchange (malonate) would not favour the model of twin carriers having identical dicarboxylate half carriers. However, a modification of the half carriers due to the influence of their partners can be visualized.

Particularly useful would be specific inhibitors able to block selectively single carriers. Until now, however, the results obtained with inhibitors have been difficult to interpret and, judging from the literature, even contradictory. Butyl malonate as an inhibitor of the malate–P_i exchange should also inhibit homo-exchanges catalysed by this carrier. However, the malate–malonate (malate) exchange does not appear to be inhibited by butyl malonate.[41, 42] This can be explained by homo-exchanges catalysed by the ketoglutarate carrier.[42] The P_i–P_i exchange should also be catalysed both by P_i and dicarboxylate carriers. In agreement with this scheme, the SH reagent N-ethylmaleimide (NEM) inhibits the P_i–P_i exchange only partially, but completely when butyl malonate is also present.[42, 45] It is not clear whether SH reagents react with the P_i carrier exclusively. In contrast to NEM, mercurials alone fully inhibit the butyl malonate-sensitive P_i–P_i exchange and the malate–P_i exchange at higher concentrations.[41, 42] At concentrations just sufficient to inhibit the butyl malonate-insensitive P_i–P_i exchange, the malate–P_i

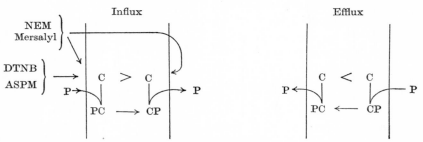

FIG. 7. Differential effects of SH reagents on inner and outer P_i carrier sites, indicating different steady state carrier distribution towards inner or outer side during influx $(C_e > C_i)$ and efflux $(C_e < C_i)$.

exchange is not affected. Consequently, the evidence afforded by inhibitors on the carrier mechanism tends to eliminate the twin carrier model.

Another aspect of the carrier mechanism to be investigated is the varying distribution of binding sites at the inner or outer membrane surfaces. By applying permeable or impermeable SH reagents, different degrees of inhibition of the P_i transport can be obtained which may be interpreted as reflecting a varying localization of the carrier to the outer or inner site[46] (cf. Fig. 7). Thus, impermeable SH reagents (ASPM, DTNB) inhibit the influx of P_i into the mitochondria, whereas the efflux of P_i generated from ATP hydrolysed in the mitochondria is not inhibited by these reagents but only by NEM and mercurials. An inhibition also of the efflux by DTNB is obtained when the carrier is apparently redistributed to the outer surface by addition of exogenous P_i.

It is hoped that the application of differently permeable inhibitors may be a fruitful way to gain further insight into the mechanism of these carriers. The interaction of various carrier systems complicates these studies, but also presents a challenge to understand this amazingly organized and intricate metabolite transport system.

F. KINETICS OF METABOLITE TRANSLOCATION

Quantitative data on the kinetics of translocation of metabolites such as P_i and dicarboxylates are rather meagre and preliminary. The equilibration of the added metabolite with the intramitochondrial space is so rapid that most separation techniques are not sufficiently fast, since time resolutions of about 2 s are required even at low temperature. Indirect measurements based upon turbidity changes could, in principle, by appropriate calibrations serve to record the net movements of the metabolites. The application of the inhibitor stop method, successfully used for the AdN translocation, to the metabolite transport is somewhat limited by the rather low effectiveness of the inhibitors. Thus, inhibitors which are substrate analogues such as butylmalonate, iodobenzyl malonate, etc. must be applied in rather high concentrations to stop the penetration rapidly. With this method for succinate at 10° C, values of $V_{Max} = 50$ μmol/min per g protein and $K_m = 0.8$ mM are obtained,[9] and for malate $K_m = 0.25$ mM.[47] For the P_i uptake SH reagents are effective inhibitors even at low concentrations. However, the reaction time with inhibitors, even for mercurials, is close to 1 s. Using mersalyl as the inhibitor for stopping the transport, $V = 20$ μmol/min per g protein (at 0° C) was measured for P_i.[42] Further progress in the kinetics of metabolite transport is urgently needed for a quantitative understanding of their physiological importance and for an analysis of the catalytic mechanism.

V. Group Translocation

Several cases of cellular metabolite transport are known where the substrate undergoes chemical modifications associated with its translocation. The modification may be only intermediate, so that as a net result the unchanged substrate appears on both sides of the membrane, or the modified substance is released to the other phase. The overall reaction may be schematized simply as follows, where only groups G and G′ are transferred:

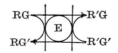

In principle, G could also be a very simple group, such as H in examples without counter transport where R alone returns after release of the transported group G' to the other side.

An important aspect of group translocation is its great potential in resolving questions of carrier-driven transport. An experimental advantage of a group translocator compared with a substrate translocator (carrier) is the possession of enzymic activity, which may persist even after the membrane has been disintegrated and can therefore be used as a convenient assay during isolation.

In the mitochondrial membrane a number of reactions linked to dehydrogenation of substrates, to electron transport and to ATP synthesis, may eventually be categorized as group translocations, when more is known about the mechanism. In the present context two metabolite transport systems will be discussed, that for aspartate and that for fatty acids, which may be regarded as group translocations.

Aspartate does not penetrate into the mitochondria, as is shown both by swelling studies and direct measurements.[44] However, intramitochondrially [^{14}C]oxaloacetate and [^{14}C]malate appear when [^{14}C]aspartate is added. This reaction is stimulated by the addition of ketoglutarate and glutamate which, in principle, have their own carrier systems. It is visualized that possibly a membrane-bound glutamate-aspartate amino transferase (transaminase) catalyses the transport of the deaminoaspartate moieties through the membrane, forming oxaloacetate on the inside.[44] The aminated transaminase reacts with ketoglutarate to form glutamate which is released on the inside. The preliminary model needs

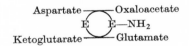

to explain the catalysis by ketoglutarate–glutamate and the inhibition of aspartate entry by hydroxylamine, which reacts with the pyridoxal phosphate of the transaminase.

The entry of fatty acids through the intramitochondrial membrane can also be regarded as a group translocation, although its mechanism is not yet clear, despite extensive efforts. A detailed discussion of the problem involving the role of carnitine has already been presented in these Essays.[48] It is generally agreed that one role of carnitine is the transfer through the mitochondrial membrane of acyl groups which can originate from external acylcarnitine or acyl-CoA. An attempt to rationalize this role of carnitine in terms of a group translocation is given in the following scheme.

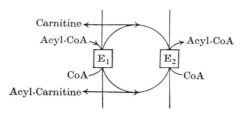

All acyl-transferase appears to be located on the inner membrane of the mitochondria.[49, 50] Part of the transferase, located on the outer surface, equilibrates with the external acyl-CoA and acylcarnitine. Another portion of the transferase, located more towards the inner phase, reacts with the endogenous acyl-CoA and with the acylcarnitine and carnitine coming from the outside. Carnitine and its derivative are unable to enter the inner phase.[51] Therefore they reach the enzyme only from the outside and it is at this stage that the actual translocation of acyl-groups from the carnitine to CoA takes place. Some evidence that there are two differently reacting transferases comes from the use of bromoacyl-CoA, which inhibits only the external step, the formation of acylcarnitine from exogenous acyl-CoA, but not the second step, the formation of inner acyl-CoA from external acylcarnitine.[52] Besides this differential inhibition, differential extraction studies also indicate the existence of two enzyme pools, one being more loosely and the other more tightly bound.[53, 54]

VI. Regulation of Metabolite Transport

The nature of the energy source which allows the metabolites to be distributed unevenly across the mitochondrial membrane has been the subject of some debate. The distribution depends on the dissociation state of the metabolite which is transported through the membrane (cf. Fig. 8). The extent of transport of anions will be determined by the membrane potential (eqns. (1) and (2)), and that of undissociated acids by the ΔpH (eqns. (3) and (4)). For the case of transport of acids the measured distribution $A_e^+ < A_i^+$ requires that $pH_i > pH_e$. Actually, in mitochondria the ΔpH is assumed to be linked to a high membrane potential which is negative on the inside (i.e. opposite to the first case), generated by the H^+-pump of the respiratory chain.[55]

The metabolites have also been considered to be transported as anions, since they are dissociated at neutral pH. In connexion with cation

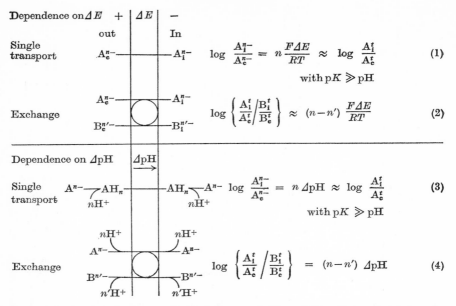

Dependence on ΔE $\quad+$ $\quad|\Delta E|$ $\quad-$

out $\quad\quad$ In

Single transport $\quad$ A_e^{n-}————A_i^{n-} $\quad \log \dfrac{A_i^{n-}}{A_e^{n-}} = n\dfrac{F\Delta E}{RT} \approx \log \dfrac{A_i^t}{A_e^t}$ $\quad$ (1)

$\quad\quad\quad\quad\quad\quad\quad\quad\quad\quad\quad\quad\quad\quad\quad\quad\quad\quad$ with $pK \gg pH$

Exchange $\quad$ A_e^{n-}———A_i^{n-} $\quad \log \left\{ \dfrac{A_i^t}{A_e^t} \Big/ \dfrac{B_i^t}{B_e^t} \right\} \approx (n-n')\dfrac{F\Delta E}{RT}$ $\quad$ (2)

$\quad\quad\quad\quad$ $B_e^{n'-}$———$B_i^{n'-}$

Dependence on ΔpH $\quad|\Delta pH \atop \longrightarrow|$

Single transport $\quad$ $A^{n-}\!\!-\!\!\nearrow\!AH_n$————$AH_n\!\!\nwarrow\!\!-A^{n-}$ $\quad \log \dfrac{A_i^{n-}}{A_e^{n-}} = n\,\Delta pH \approx \log \dfrac{A_i^t}{A_e^t}$ $\quad$ (3)

$\quad\quad\quad\quad$ $nH^+ \quad\quad\quad\quad\quad\quad nH^+$ $\quad\quad\quad\quad\quad\quad\quad\quad\quad$ with $pK \gg pH$

$\quad\quad\quad\quad$ $nH^+ \quad\quad\quad nH^+$

Exchange $\quad$ $A^{n-}\!\!\smile\!\!\!\!\frown\!\!\!\!A^{n-}$ $\quad \log \left\{ \dfrac{A_i^t}{A_e^t} \Big/ \dfrac{B_i^t}{B_e^t} \right\} = (n-n')\,\Delta pH$ $\quad$ (4)

$\quad\quad\quad\quad$ $B^{n'-}\!\!\smile\!\!\!\!\frown\!\!\!\!B^{n'-}$

$\quad\quad\quad\quad$ $n'H^+ \quad\quad n'H^+$

Fɪɢ. 8. Distribution of anions under the influence of ΔE or ΔpH. e, External; i, internal; n-, n-', anionic charges; A^t, total substrate concentration.

movements, the anions were thought to follow the actively transported cations creating a membrane potential which is positive on the inside.[56]

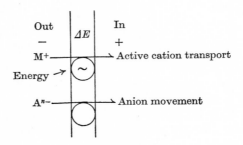

Out $\quad\quad|\Delta E|\quad\quad$ In

$-\quad\quad\quad\quad\quad\quad +$

M^+———→ Active cation transport

Energy ↗ (~)

A^{n-}———→ Anion movement

From the equilibrium distribution of anions according to the Nernst equation, a membrane potential $\Delta E \sim 20$ to 30 mV positive inside has been calculated.[57]

This interpretation of the metabolite transport is contested by considerable experimental evidence. First it requires that cations are transported by an energy-driven carrier against this postulated electric gradient. However, this is contrary to the widely accepted assumption that ionophores catalyse passive cation transport following an electric potential gradient which should be negative inside for a cation accumulation.

Evidence for the mode of penetration of the metabolites as "acids" is forthcoming from (a) measurements of metabolite distribution dependent on ΔpH,[58] (b) the movements of H^+ accompanying metabolite transport, and (c) the dependence of anion permeability on the conductivity of the membranes for cations or H^+.[59]

There is little doubt that the monocarboxylic acids can diffuse through the membrane rapidly without a carrier, as discussed above, because they can maintain a sufficiently high concentration of undissociated acid in the membrane. This is favoured by the lower dielectric constant of the lipid phase and the more lipophilic properties of the undissociated acid.

The concentration of di- and tribasic acids in the membrane phase cannot be expected to reach levels sufficient for rapid metabolite transport, since in the biphasic distribution the equilibrium concentration of undissociated acid in the aqueous phase is too low, being 10^{-5} to 10^{-7}. Therefore these metabolites need a carrier (C) on whose binding site the anion may be bound to a basic group, which dissociates on either side of the membrane together with the anions (cf. also Ref. 60):

$$A^{2-} + 2H^+ \quad \longrightarrow \quad \overset{C}{\underset{}{|}} \quad AH_2C \quad \overset{C}{\underset{}{|}} \quad \longrightarrow \quad A^{2-} + 2H^+$$

where AH_2C is the translocation complex of H^+, carrier and dicarboxylate. There may be first binding of H^+ to a basic carrier group and the following attachment of the anion:

$$C + 2H^+ \rightarrow CH^{2+} + A^{2-} \rightarrow CH_2A$$

or binding of first the anion to the carrier and then binding of H^+:

$$C + A^{2-} \rightarrow CA^{2-} + 2H^+ \rightarrow CAH_2$$

Evidence for the penetration of metabolites as acids coupled to H^+ movement is provided by studies on the effect of conductors for K^+ and H^+ on anion movements.[12, 59] Thus the movements of anions are shown to require an efflux of H^+, in exchange for cations, which reflects the H^+ accumulated together with the metabolites. Furthermore, it was observed that in the presence of dicarboxylic substrates H^+ is more easily conducted into the membrane, obviously carried by the metabolites.[61] Recently the relation of the H^+ movement to the uptake of metabolites has been measured and ratios $H^+/P_i \leqslant 1\cdot8$ and $H^+/malate = 2$

have been found when corrected for exchange between anions.[42] The exchange between dicarboxylates and P_i is not accompanied by any H^+ movement. Only in the exchange citrate $\leftrightharpoons$ malate can a H^+ uptake be observed: $H^+/$citrate–malate $= 0.9$.

Of particular importance are the studies on the distribution of metabolites across the membrane and their dependence on ΔpH. Thus measurements of the ΔpH and the metabolite distribution at various substrate concentrations reveal good agreement with eq. (3) of Fig. 8. The ratio $(P_i)_i/(P_i)_e$, as measured in relation to ΔpH, is found to follow the slope $n \approx 1.5$, which corresponds to the degree of dissociation around pH $7.$[58] The comparison of dependence on ΔpH of mono-, di- and tricarboxylates in their distribution across the membrane impressively supports this relation. At low external concentrations the corresponding ratio $(A_i)_t/(A_e)_t$ follows $n \approx 1$ for acetate, $n \approx 2$ for malate and $n \approx 3$ for citrate, in good agreement with the theory that the distribution is dependent on parallel movements of H^+ or transport of undissociated acids.[58]

One would expect the transport of P_i to be compensated by H^+ movement, since there is no exchange against another anion which could compensate a charge movement where P_i is transported as an anion. In fact, in order to maintain the exchange principle, the P_i transport has been visualized as an exchange against OH^-, which has been written as:

This particular exchange is barely distinguishable from the formulation of an electroneutral transport of the undissociated acids, as written above. The present evidence cannot eliminate the possibility that the dicarboxylates and tricarboxylates are, in fact, transported as anions in exchange reactions, e.g. malate$^{2-} \rightleftharpoons P^{2-}$ and citrate$^{2-} \rightleftharpoons$ malate^{2-}. Since these metabolites are coupled ultimately to the transport of P_i, both the observed H^+ movement and the pH dependent distribution of di- or tricarboxylates can be considered to be a result of the H^+-coupled cycling P_i shuttle (Fig. 4).

In summary, the same pH dependent distribution and the same H^+ movement, accompanying those metabolites which are coupled to P_i transport can be expected whether they are transported by the exchange carriers as A^{n-} or as AH_n. However, the arguments on the mechanism of carrier transport quoted above make it more probable that the metabolites are exchanged by the carriers as AH_n. At least in one case, independent of P_i transport, in the exchange citrate $\rightleftharpoons$ malate, the accompanying transport of H^+ is evidence for a compensation by H^+ of one carboxylate group of the citrate.

In the light of this interpretation of metabolite transport, the massive movements of anions accompanying the active cation (K^+, Ca^{2+}) accumulation can be explained as follows. Cations move into the membrane following a potential gradient. The disturbance of the electrochemical equilibrium activates the H^+-pump of the respiratory chain, which leads to increased H^+ efflux. As a result the inner pH is raised and, by an influx, the metabolites are redistributed.

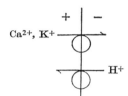

H^+-pump (electron transport)

$$A^{n-} + nH^+ \longrightarrow AH_n$$

$(H^+)_e > (H^+)_i$

VII. The Transport of Adenine Nucleotides

A. GENERAL

The transport of adenine nucleotides is a particularly important function of the mitochondrial membrane since the supply of ATP to the extramitochondrial space is the major task of the mitochondria. In quantitative terms the AdN transport would also be expected to be important: for the oxidation of 1 mol pyruvate up to 15 mol ADP and ATP are taken up or released by the mitochondria. An amount of P_i equal to ADP would be expected to be transported into the mitochondria. However, no coupling between the transport of AdN and P_i has been found. This appears to be reasonable in view of the coupling between transport of P_i to di- and tricarboxylates.

The relatively large size of the AdN molecules, their hydrophilic properties and the high degree of ionization require catalysis of its membrane transport by a carrier. This is confirmed by the extensive evidence about the AdN translocation, which is the best characterized transport process in mitochondria. In these studies new techniques for the direct determination of metabolite transport in mitochondria were developed. The use of three different isotopes (^{14}C, ^{3}H and ^{32}P) for labelling AdN has made possible various experiments with double labelling, such as the study of balance of release and uptake, changes in phosphorylation pattern, competition between ADP and ATP, etc.

A wide range of properties for this exchange has been described: 1, specificity of transport in both directions; 2, quantitative analysis of the kinetics and kinetic parameters, and of temperature dependence; 3, regulation of AdN translocation; 4, quantitative correlation of the AdN translocation with oxidative phosphorylation; 5, quantitative elucidation of carrier sites and their properties on the membrane; 6, qualitative and quantitative correlation of transport parameters with extramitochondrially measured functions.

B. PRINCIPAL FEATURES OF AdN TRANSLOCATION AND SPECIFICITY (Table 3)

The AdN translocation was early characterized as a 1:1 exchange between exogenous and endogenous AdN.[24, 62] Most studies on this translocation have dealt with liver mitochondria, which contain endogenous AdN in relatively high amounts. As a consequence of the exchange, the intramitochondrial AdN pool remains constant even at relatively low external AdN concentrations. The AdN exchange is highly specific for ADP and ATP.[2] AMP is not transported, so that the intra- and extramitochondrial AMP remain separated. This has important metabolic consequences, mentioned here only briefly: the intra- and extramitochondrial AMP depots are rephosphorylated to ADP by separate systems; within the mitochondrial matrix by the GTP-AMP-transphosphorylase coupled to the substrate level phosphorylation, and outside by the adenylate kinase which is exclusively located in the intramembrane space.

It was demonstrated by double labelling of [^{14}C, γ-^{32}P]ATP that ATP is transported as an intact molecule and that there is no transphosphorylation between intra- and extramitochondrial AdN.[2, 24] The highest uptake of [γ-^{32}P]ATP was found in the uncoupled state—an apparent paradox, to be discussed below under "Regulation of AdN Exchange".

Of great interest is the fact that there are highly effective and specific inhibitors of the AdN translocase: atractyloside and bongkrekic acid (cf. Table 4). Atractyloside has long been known as a potent inhibitor of oxidative phosphorylation (cf. Ref. 63) but its mechanism of action could only be resolved after definition of the AdN translocation was achieved.[24, 62] The inhibition by atractyloside seems to be competitive with AdN and that by bongkrekic acid to be non-competitive. Therefore

TABLE 3

Summary of data on adenine nucleotide translocation

Specificity	Base: Adenine (no other common base is active) ribose (5–10×) as active as deoxyribose Phosphate: 5' position, $P_n (n = 2$ to 4) (no activity, 3', 2' positions, $n = 1$) ADP = ATP and (10–20×) as active as A-tetra-P
Inhibitors	Atractyloside $K_i \approx 10^{-7}$ M, competitive Bongkrekic acid $K_i \approx 2 \times 10^{-8}$ M, non-competitive
Kinetics	1 mol : 1 mol exchange First-order equilibration of endogenous [^{14}C]AdN with exogenous AdN Zero-order (maximum) rate = first order rate × conc. endog. (ADP + ATP), (μmol/min per g protein) For ADP: at 0° C, $V_T = 7$ at 20° C, $V_T \approx 200$ } μmol/min per g protein $K_m = 12$ μM Temp.–dependence: $E_A = 34$ (0–8° C), 21 (8–20° C) kcal

Carrier turnover, see Table 4; Regulation, see Table 6.
V_T, Translocation activity; E_A, energy of activation/mol.

TABLE 4

Reaction with specific inhibitors

Inhibitor	Atractyloside	Bongkrekic acid
Type of inhibition	Competitive	Uncompetitive (cooperative?)
K_i (estimated)	$\approx 2 \times 10^{-8}$ M	$\approx 2 \times 10^{-8}$ M
Number of inhibition sites, for liver mitochondria (μmol per g protein)	≈ 0.3 to 0.4	0.2
Effect on carrier	Removes AdN	Fixes AdN
Reactivity measured both on exchange $t_{1/2}$ and on carrier	≈ 20 ms at 25° C < 1 s at 0° C	≈ 30 s at 25° C Nonreactive at 0° C

at high AdN concentration, bongkrekic acid is a still more effective inhibitor of the translocation than is atractyloside.[25, 26] The principal difference in the action of the two inhibitors on the AdN carrier will be discussed below.

C. KINETICS OF AdN TRANSLOCATION

A quantitative understanding of the kinetics of the AdN exchange was possible after it was shown that the inside–outside reaction of the exchange also excludes AMP.[2] The exchangeable part of the intramitochondrial AdN pool is therefore confined to the ADP + ATP portion. The total pool is functionally compartmented for the exchange into the ADP + ATP and the AMP portions, which can vary widely under different metabolic conditions. The exchange can be quantitatively evaluated as a first order reaction if referred to the endogenous ADP + ATP pool only. True translocation rates (μmol/min per g protein), which are independent of the variable ADP + ATP pool, can be obtained

TABLE 5

Kinetic data of adenine nucleotide carrier (rat liver mitochondria)

	0° C	20° C	37° C (extrapolated)
Number of sites (μmol per g protein)	0·13		
Membrane area per site	4×10^4 (Å^2/molecule) = 250 (cm^2/μmol)		
Activity of translocation (μmol/min per g protein)	7	200	1200
Turnover per site (min^{-1})	55	1600	10000
Flux rate per membrane area (μmol/min per cm^2)	0·2	7	40

by multiplying the first-order rate by the concentration of intramitochondrial ADP + ATP. The various kinetic parameters are summarized in Table 3.

The determination of the number of carrier sites, discussed below, permits the evaluation of the turnover number for the carrier, as shown in Table 5.[64] The turnover number of this carrier appears to be well within the range found for other carriers, such as those for Na$^+$, K$^+$ and for glucose in plasma membranes. The density of occupation of the inner membrane by carrier sites and the flux rate per membrane surface can be estimated.

D. REGULATION OF AdN EXCHANGE

In the controlled state (coupled state) the exogenous ATP exchange rate is considerably slower than that of ADP.[2, 24] (Fig. 9, Table 6)

Under the influence of uncoupler or ionophores plus K^+ the exchange rates for ATP and ADP become equal.[2, 65] The difference in the translocation activities becomes more apparent under competitive conditions, when both ADP and ATP are supplied simultaneously from the outside.

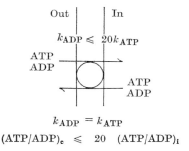

$$k_{ADP} = k_{ATP}$$
$$(ATP/ADP)_e \leqslant 20 \ (ATP/ADP)_i$$

Fig. 9. The asymmetry of the adenine nucleotide exchange.

At a ratio ADP:ATP = 1, ADP can exchange 10 times faster than ATP.[2, 65] Competition is best expressed in the apparent ratios of the K_m thus: $K_m^{ATP}/K_m^{ADP} \approx 10$. In the presence of uncoupler the competition is abolished. For the inside → outside direction there is always a competition between ADP and ATP, since both are present to varying degrees in the endogenous pool. However, in this direction ADP and ATP are equally active, independent of the controlled or uncoupled state.[2] This was demonstrated by the replacement at equal rates of both endogenous ADP and ATP with exogenous [^{14}C]AdN.

TABLE 6

Control properties of adenine nucleotide exchange

State		V_{ATP}/V_{ADP} Out→in	Competition (ADP:ATP = 1:1)
Controlled	0·2–0·5		0.05–0·1
Uncoupled	1 –1·5		1
K^+-Permeable	1 –1·5		1
Anaerobic	0·3–0·7		
		In→out	
	1		1
		K_m^{ATP}/K_m^{ADP}	
Controlled			3–10
Uncoupled			0·8–1

It is clear that asymmetry in the specificity of the AdN exchange cannot be attributed to the catalysing carrier *per se*, but must be caused by a superimposed energy source.[2, 66] This is in agreement with the fact that uncoupling of oxidative phosphorylation, which inhibits energy transfer, abolishes the asymmetry. This asymmetry can be regarded as important in the regulation of the AdN translocation, with major consequences for the phosphorylation potential of exogenous ATP. A control of the AdN translocation by the membrane potential has been considered.[2, 67] This interpretation is based on the fact that

Exchange with charge transfer (electrogenic), 1 e⁻ net movement

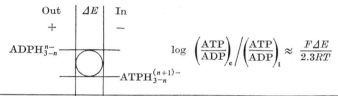

$$\log \left(\frac{ATP}{ADP}\right)_e \bigg/ \left(\frac{ATP}{ADP}\right)_i \approx \frac{F\Delta E}{2.3RT}$$

Exchange with H⁺ transfer (electroneutral), 1 H⁺ net movement

$$\log \left\{\left(\frac{ATP}{ADP}\right)_e \bigg/ \left(\frac{ATP}{ADP}\right)_i\right\} \approx (H^+)_i - (H^+)_e$$

Sum of influence of ΔE and ΔpH when fraction α is transported electrogenically

$$\log \left\{\left(\frac{ATP}{ADP}\right)_e \bigg/ \left(\frac{ATP}{ADP}\right)_i\right\} \approx \alpha \frac{F\Delta E}{2.3RT} + (1-\alpha)\left\{(H^+)_i - (H^+)_e\right\}$$

Fɪɢ. 10. Distribution of ATP/ADP across the membrane under the influence of ΔE and ΔpH.

increased conductance of the membranes for H⁺ (by uncouplers) or for K⁺ largely abolishes the asymmetry. In further studies the stoicheiometric relation of H⁺ and K⁺ movements to that of AdN were measured.[68] In the presence of a K⁺ conductor 0·4 K⁺/ATP, and in the presence of a H⁺ conductor (uncoupler) 0·6 H⁺/ATP, are taken up. In the reverse reaction ADP ⇌ ATP about 0·4 H⁺/ATP are released.[65] It is concluded that the ATP ⇌ ADP exchange is partially H⁺-compensated and electroneutral, and partially electrogenic. The inhibition of the ATP entry is explained by the partially electrogenic reaction, assuming a membrane potential which is inside-negative and therefore would repel the excess negative charge (see Figs. 9, 10). The electroneutral portion in the

ATP $\rightleftharpoons$ ADP exchange should have an opposite effect, i.e. ATP should be attracted more to the inside than ADP. This is over-compensated by the stronger electrogenic repulsion.

Of great significance are the shifts of the ratio ATP/ADP between the inner and outer space as a result of these membrane forces. The membrane potential will, as indicated, increase the ratio ATP/ADP outside versus inside. The pH, increasing from outside to inside, would have the opposite effect. The actual distribution is a sum of both effects, i.e. those on the electrogenic and electroneutral portions of the exchange. In general (cf. Ref. 55) the electromotive force of the membrane potential would be considerably larger than the chemical force of the pH and therefore it would mainly determine the distribution of ATP/ADP. Since the ratio ATP/ADP is higher inside than outside, one may surmise that the predominant control is by the membrane potential and not by the ΔpH. Here the AdN translocation differs in an important regulatory effect from the transport of the other metabolites discussed above.

Particular consequences result from this regulation for the phosphorylation potential of ATP. Here one has to take into account also the distribution of P_i which is controlled mainly by pH:

$$\log (P_i)_i/(P_i)_e = 1.8(pH_i - pH_e) \text{ at pH } 7$$

The pH difference across the membrane should not shift the phosphorylation potential from the inside to the outside since the ΔpH-dependent distribution of the components represents an equilibrium which can be expressed as follows:

$$G = G_H^o + 2 \cdot 3 RT \left\{ \log \frac{ATP_o}{ADP_o \times P_{io}} - \log \frac{\varphi_{ATP}}{\varphi_{ADP} \times \varphi_{P_i}} \right\}$$

where G_H^o is the free enthalpy of undissociated forms

$$\varphi = \left(1 + \frac{K_1}{[H^+]} + \frac{K_1 \times K_2}{[H^+]^2} + \cdots \cdots \right)$$

K_n are dissociation constants, ATP_o, ADP_o, P_{io}, are total concentrations, and, from $G_e = G_i$,

$$\Delta_{e-i} \log \frac{ATP}{ADP \times P_i} = \Delta_{e-i} \log \frac{\varphi_{ATP}}{\varphi_{ADP} \times \varphi_{P_i}}$$

The same relation would be obtained by equalizing the inner and outer concentrations of the undissociated forms:

$$(ATPH_4)_i = (ATPH_4)_e, \text{ etc.}$$

TABLE 7

Comparison of various calculated and measured inner (i) and outer (e)
ATP-potentials

1. *From ΔE and ΔpH* (at 25°C)

   Assumption: $\Delta E = 200$ mV, $\quad \Delta pH = -0.5$ (cf. Refs. 55, 58)

   $\qquad \alpha = 0.5, \qquad\qquad n' = 1.8$ for P_i

   $$\Delta_{e-i} \log \left(\frac{ATP}{ADP \times P_i} \right) = \alpha \left(\frac{\Delta E}{60} - n'\Delta pH \right) = 2.1$$

   $\Delta G_{e-i} = 1.39 \times 2.1 = 2.9$ kcal

2. *From translocation ratios*

   $$\left(\frac{ATP}{ADP} \right)_e \Big/ \left(\frac{ATP}{ADP} \right)_i = \frac{\vec{k_D}}{\vec{k_T}}$$

   $$\Delta G_{e-i} = 1.39 \left\{ \log \left(\frac{ATP}{ADP} \right)_e \Big/ \left(\frac{ATP}{ADP} \right)_i - \log (P_i)_e/(P_i)_i \right\}$$

   $$= 1.39 \left\{ \log \frac{\vec{k_D}}{\vec{k_T}} - \Delta \log P_i \right\}$$

   Assumption: $(P_i)_e/(P_i)_i = 0.1, \qquad \dfrac{\vec{k_D}}{\vec{k_T}} = 20$

   $\Delta G_{e-i} = 3.2$ kcal

3. *From P/O difference*

   $$\Delta G_{e-i} = \frac{(P/O)_{max} - (P/O)}{(P/O)_{max}} \Delta G_e = \gamma \Delta G_e$$

   $\Delta G_t = \Delta G_e = 12.7$ kcal $\qquad \gamma = 0.26$

   $\Delta G_{e-i} = 3.4$ kcal

4. *From contents of ATP, ADP, P_i*

	$\dfrac{ATP}{ADP}$	P_i (μM)	ΔG (kcal)
Intramitochondrial	3.2	16.2	12.0
Extramitochondrial	20.2	1.0	14.8

   $\Delta G_{e-i} = 2.8$ kcal

$\vec{k}$ = Rate constant out $\rightarrow$ in; α = fraction transferred electrogenically.

A difference in the phosphorylation potential across the membrane is therefore generated only by that fraction α of the AdN which is driven by the membrane potential. The ΔpH further increases this difference by its effect on the P_i distribution which is assumed to follow only the pH difference. From this relation the phosphorylation potential difference or translocation energy $\Delta G = 3.2$ kcal is calculated, assuming simple values approximating to the measured data (see Table 7). For comparison the ΔG is estimated also from the shift of ATP/ADP which should result from the ratio of the rates for ATP and ADP entry, $k_D^{\rightarrow}/k_T^{\rightarrow}$. There is approximate agreement between the measured ΔG and those calculated from the membrane potential, and the translocation rates.

The energy necessary to translocate ATP against the increased phosphorylation potential outside is derived from the same source as for the phosphorylation of ATP. Therefore the yield in the formation of ATP, i.e. the P/O ratio, should be decreased.[68] This P/O deficit increases with the exogenous ATP/ADP[69] and can also be used to calculate ΔG, as shown in Table 7.

As a result about 20–30% of the total phosphorylation energy of ATP is obtained through the translocation and only 70–80% by way of the ATP synthesis *per se* in the mitochondria. The apparently paradoxical result of this energy-dependent translocation is the lowering of the ATP potential where ATP is formed and increasing it in the compartment where it is consumed.

VIII. The Carrier for Adenine Nucleotides and its Properties

The elucidation of the molecular foundation of metabolite transport is only in its early stage. Until such time as it is possible to isolate a carrier and to deduce its function and properties from its molecular structure, the carrier can be studied *in situ* in the mitochondrial membrane at the molecular level. Thus the number of carrier sites in the membranes should be determined in order to know their molecular abundance. Once the binding of substrates to the carrier can be measured, a number of important characteristics of the carrier can be elucidated, such as the dissociation constant, the specificity of the binding, influence of parameters such as pH, temperature, metal ions, etc. The substrate binding can be used for elucidating the inhibitor mechanism.

Modified substrates or inhibitors can be devised as environmental indicators containing fluorescent or paramagnetic groups. The binding to the carrier *in situ* might reveal the environmental change of the carrier site which accompanies the actual translocation of the substrate through the membrane. Reagents may be used to identify amino acids at the active binding site. From here a bridge leads to the thorny path of isolating the carrier. With a specific and covalent label at the binding site, rather drastic isolation procedures can be applied for separating the hydrophobic protein from the lipid environment, whereas the substrate binding cannot be expected to survive this procedure.

For the AdN carrier a small part of this programme has been success-fully pursued. The characterization of the AdN carrier is facilitated by the relatively high affinity for AdN ($K_D = 10^{-6}$ M) and the existence of a highly effective inhibitor (atractyloside) which specifically removes the bound AdN.[64, 70] On this basis the carrier-bound AdN could be discriminated from unspecifically bound or trapped AdN, as well as from the exchanged AdN. Since binding to the carrier sites can be ex-pected to represent only a few percent of the AdN exchanged with the endogenous pool, the mitochondria have to be depleted as far as possible of endogenous AdN. On this basis a sensitive discriminating procedure

TABLE 8

Summary of properties of adenine nucleotide carrier[70]

Mitochondria source	Rat liver ADP	Rat heart ADP	Beef heart	
			ADP	ATP
Sites/cyt. a (mol/mol)	1·2	2·2	1·3	1·4
Types of sites (%)				
High:low affinity	100:0	25:75	20:80	17:83
Dissociation constant	0·5	0·1	0·1–0·3	0·6
(μM)	—	4·1	4–7	12
Specificity	Binding of ADP=ATP, no binding of AMP or non-adenine nucleotides			
Temperature dependence	Very small change of K_D between 0–25°			
pH dependence	80% decrease of site number from pH 7·0 to pH 8·0; pK = 7·2			
Cation^{2+} dependence	Decrease of affinity by Ca^{2+} and Mg^{2+}			
Group reagents	Sensitive to histidine reagents			

for the elucidation of the carrier site was worked out, which permits accurate quantitative studies.

Interestingly, the number of carrier sites in the membrane is approxi-mately equal to the number of cytochrome oxidase molecules (cf. Table 8). This ratio is rather constant, even when the cytochrome content per unit of protein changes fourfold, as it does from rat liver to beef heart. The relatively great number of sites reflects the prime importance of the carrier for oxidative phosphorylation.

The specificity of binding to the carrier is similar to that of the AdN exchange. Only ADP and ATP, and ADP-N-P, ADP-C-P analogues (not AMP) bind with high affinity. A number of other properties of the carrier binding also agree with those predicted by the translocation. Of particular interest is the concentration-dependence of the binding. A clear saturation of the carrier binding at low concentrations of AdN was found, in contrast to the unspecific binding. The concentration dependence in a Scatchard plot revealed single binding sites with high affinity for liver mitochondria, and a non-linear relation for heart mitochondria, representing at least two binding sites with high and low affinities. The occurrence of two different affinities is interpreted to reflect inside and outside localized carrier sites.[70, 71] In terms of a mobile carrier system it is visualized that about one-quarter of the carrier sites are turned to the inner site, where they are nearly saturated with endogenous AdN. Thus even at low concentrations of exogenous [14C]AdN the inner concentration of [14C]ADP becomes saturating by exchange with the endogenous AdN. The inner sites therefore have an apparently small K_D, and the outer sites have the higher true K_D values, where K_D is the dissociation constant for ADP. This interpretation implies a common pool of inner and outer sites with a single affinity, corresponding to the higher K_D. The double carrier model has also been considered for the AdN exchange, but here a 1:1 ratio of inner to outer sites would be required.

The pH dependence ($pK = 7.2$) and the inhibition of carrier binding by methylene blue catalysed photo-oxidation indicate that histidine is linked to the binding site. It can well be expected that a positively charged group such as histidine, possibly together with a tightly bound divalent cation, contributes to the binding. From the specificity of translocation for di- and triphosphate nucleotides, it was early concluded that the substrate of the AdN carrier should have at least three anionic groups. This rule extends to the specific inhibitors atractyloside and bongkrekic acid, both being tribasic acids.

The binding to the AdN carrier is strongly influenced by the inhibitors of the AdN translocation (cf. Table 4). By the very definition of the specific carrier binding sites, the atractyloside-removability, it is clear that atractyloside removes AdN competitively from the carrier site. The expected 1:1 stoichiometry for atractyloside binding is not yet clearly established. From the concentration dependence of ADP removal by atractyloside, a $K_D \approx 10^{-7}$ to 10^{-8} M can be estimated,[70] in agreement with the constant for the inhibition of translocation $K_i \approx 10^{-8}$ M.[65] Thus the affinity of the binding site for atractyloside must be 50 to 100 times higher than for AdN.

The effect of bongkrekic acid on the carrier is opposite to that of atractyloside: the affinity of binding of AdN is increased more than 100-fold, so that for ADP the binding constant $K_D < 10^{-8}$ M.[72] As a result the dissociation of ADP is so slow that exchange of the

[^{14}C]ADP with cold ADP needs much longer ($t_{1/2} \approx 5$ min) than with the ADP bound to untreated sites ($t_{1/2} \ll 1$ s). Moreover, atractyloside does not remove the tightly bound ADP. This, and the results of other experiments, indicate that bongkrekic acid can increase the affinity of the carrier only specifically to ADP but not to atractyloside. This appears to be reciprocal, since the binding of bongkrekic acid is also increased by ADP.

Of particular interest was the finding that the number of binding sites for ADP is lower by about 20% when bongkrekic acid is added before ADP than when added after ADP.[72] Only in the latter case are the maximum number of carrier sites occupied, as revealed by parallel atractyloside removal experiments. The difference is equal to the number of high affinity sites found without bongkrekic acid. This provides strong independent evidence that the high affinity sites are located on the inner sides of the membrane. The endogenous AdN becomes fixed to the sites under the influence of bongkrekic acid, so that subsequent addition of [^{14}C]ADP cannot bind to the inner sites but only the empty outer sites. If [^{14}C]ADP equilibrates first with endogenous AdN, [^{14}C]-ADP is then fixed both to inner and outer sites under the influence of bongkrekic acid:

A* = [^{14}C]AdN; B = bongkrekic acid; C = carrier.

IX. Conclusions

The reader may well have noted that the field of mitochondrial metabolite transport, which is part of the function of the inner mitochondrial membrane, is still in its developmental stage and that entirely new results may be reported, demanding a major change in current views. The field can be considered to be at an exciting stage which invites more biochemists to participate in its development. However, the difficulties which face the elucidation of the molecular basis should not be overlooked. The elucidation of the carriers, their binding sites, their mechanisms and last but not least their isolation and chemical identification, meets with the same experimental problems as those encountered in the elucidation of carriers in other biological membranes. These difficulties are twofold: the frequent absence of an appropriate assay for a carrier, and the problem of dealing with hydrophobic interactions between the proteins and membrane lipids.

Nevertheless the research in this field represents a challenge which, after penetration of the "membrane barrier", might provide deep insight into the function and biological meaning of the membranes. In this respect the arduous present work in this field may one day turn out to be thoroughly rewarding.

REFERENCES

1. Klingenberg, M. (1970). Mitochondrial metabolite transport. *FEBS Lett.* **6**, 145–154.
2. Pfaff, E. & Klingenberg, M. (1968). Adenine nucleotide translocation of mitochondria. I. Specificity and control. *Eur. J. Biochem.* **6**, 66–79.
3. Pressman, B. C. (1958). Intramitochondrial nucleotides. I. Some factors affecting net interconversions of adenine nucleotides. *J. biol. Chem.* **232**, 967–978.
4. Amoore, R. E. & Bartley, W. (1958). The permeability of isolated rat liver mitochondria to sucrose, sodium chloride and potassium chloride at 0°. *Biochem. J.* **69**, 223–238.
5. Harris. E. J. & Van Dam, K. (1968). Changes of total water and sucrose space accompanying induced ion uptake or phosphate swelling of rat liver mitochondria. *Biochem. J.* **106**, 759–766.
6. Klingenberg, M., Pfaff, E. & Kröger, A. (1964). Techniques for studying kinetics in mitochondrial suspensions. In *Rapid Mixing and Sampling Techniques in Biochemistry*, pp. 333–337. Ed. by Chance, B. New York: Academic Press.
7. Klingenberg, M. & Pfaff, E. (1967). Means of terminating reactions. In *Methods in Enzymology*, vol. 10, pp. 680–684. Ed. by Colowick, H. & Kaplan, N. O. New York & London: Academic Press.
8. Kraaijenhof, R., Tsou, C. S. & Van Dam, K. (1969). The determination of the rate of uptake of substrates by rat-liver mitochondria. *Biochim. biophys. Acta*, **172**, 580.
9. Quagliariello, E., Palmieri, F., Prezioso, G. & Klingenberg, M. (1969). Kinetics of succinate uptake by rat liver mitochondria. *FEBS Lett.* **4**, 251–254.
10. Pfaff, E., Heldt, H. W. & Klingenberg, M. (1969). Adenine nucleotide translocation of mitochondria. Kinetics of the adenine nucleotide exchange. *Eur. J. Biochem.* **10**, 484–493.
11. Chappell, J. B. & Haarhoff, K. N. (1966). The penetration of the mitochondrial membrane by anions and cations. In *Biochemistry of Mitochondria*, pp. 75–91. London & Warsaw: Academic Press and Polish Scientific Publishers.
12. Mitchell, P. & Moyle, J. (1969). Translocation of some anions, cations and acids in rat liver mitochondria. *Eur. J. Biochem.* **9**, 149–155.
13. Chappell, J. B. (1968). Systems used for the transport of substrates into mitochondria. *Br. med. Bull.* **24**, 150.
13a. Harris, E. J. & Manger, J. R. (1968). Intramitochondrial substrate concentration as a factor controlling metabolism: the role of interanion competition. *Biochem. J.* **109**, 239–246.

13b. Williamson, J. R., Anderson, J. & Browning, E. T. (1970). Inhibition of gluconeogenesis by butyl malonate in perfused rat liver. *J. biol. Chem.* **245**, 1717–1726.

14. Chappell, J. B. & Robinson, B. H. (1968). Penetration of the mitochondrial membrane by tricarboxylic acid anions. In *The Metabolic Roles of Citrate*. Ed. by Goodwin, T. W. London & New York, Academic Press. *Biochem. Soc. Symp.* **27**, 123–133.

15. Palmieri, F. & Klingenberg, M. (1967). Inhibition of respiration under the control of azide uptake by mitochondria. *Eur. J. Biochem.* **1**, 439–446.

16. Rossi, C. S. (1969). Discussion remarks in *The Energy Level and Metabolic Control of Mitochondria*, pp. 74–75. Bari: Adriatica Editrice.

17. Chappell, J. B. & Crofts, A. R. (1966). Ion transport and reversible volume changes of isolated mitochondria. In *Regulation of Metabolic Processes in Mitochondria*, pp. 293–314. Amsterdam: Elsevier.

18. Holton, F. A. (1969). Presence and spatial localization of carbonic anhydrase in rat liver mitochondria. *Biochem. J.* **116**, 29P.

19. Klingenberg, M. & Buchholz, M. (1970). Localization of the glycerol-phosphate dehydrogenase in the outer phase of the mitochondrial inner membrane. *Eur. J. Biochem.* **13**, 247–252.

20. Klingenberg, M. & V. Häfen, H. (1963). Wege des Wasserstoffs in Mitochondrien. I. Die Wasserstoffubertragung von Succinat zu Acetoacetat. *Biochem. Z.* **337**, 120–145.

21. Williamson, D. H., Lund, P. & Krebs, H. A. (1967). The redox state of free nicotinamide-adenine dinucleotide in the cytoplasm and mitochondria of rat liver. *Biochem. J.* **103**, 514.

22. Scholz, R. (1968). Untersuchungen zur Redoxkompartmentierung bei der hämoglobin-frei perfundierten Rattenleber. In *Stoffwechsel der perfundierten Leber*, p. 25. Heidelberg: Springer-Verlag.

23. Stein, W. D. (1967). *The Movement of Molecules across Cell Membranes*. London & New York: Academic Press.

24. Klingenberg, M. & Pfaff, E. (1965). Structural and Functional Compartmentation in Mitochondria. In *Regulation of Metabolic Processes in Mitochondria*, p. 180, Amsterdam: Elsevier.

25. Henderson, P. J. F. & Lardy, H. A. (1970). Bongkrekic acid: an inhibitor of the adenine nucleotide translocase of mitochondria. *J. biol. Chem.* **245**, 1319–1330.

26. Klingenberg, M., Grebe, K. & Heldt, H. W. (1970). On the inhibition of the adenine nucleotide translocation by bongkrekic acid. *Biochem. biophys. Res. Commun.* **39**, 344–351.

27. Fonyo, A. (1968). Phosphate carrier of rat liver mitochondria. Its role in phosphate outflow. *Biochem. biophys. Res. Commun.* **32**, 624–628.

28. Tyler, D. D. (1968). The inhibition of phosphate entry into rat liver mitochondria by organic mercurials and by formaldehyde. *Biochem. J.* **107**, 121–123.

29. Robinson, B. H. & Chappell, J. B. (1967). The inhibition of malate, tricarboxylate and oxoglutarate entry into mitochondria by 2-n-butylmalonate. *Biochem. biophys. Res. Commun.* **28**, 249–255.

30. Robinson, B. H. & Williams, G. R. (1969). The effect on mitochondrial oxidations of inhibitors of the dicarboxylate anion transporting system. *FEBS Lett.* **5**, 301–304.

31. Robinson, B. H., personal communication.

32. McGivan, J. D. & Chappell, J. B. (1969). Avenaciolide: a specific inhibitor of glutamate transport in rat liver mitochondria. *Biochem. J.* **116**, 37P.
33. Quagliariello, E. & Palmieri, F. (1968). Control of succinate oxidation by succinate uptake in rat-liver mitochondria. *Eur. J. Biochem.* **4**, 20–27.
34. Meijer, A. J., Tager, J. M. & Van Dam, K. (1969). The movement of tricarboxylic acids across the mitochondrial membrane. In *The Energy Level and Metabolic Control in Mitochondria*, pp. 147–157. Bari: Adriatica Editrice.
35. Van Dam, K. & Tsou, C. S. (1968). Accumulation of substrates by mitochondria. *Biochim. biophys. Acta*, **162**, 301–309.
36. Papa, S., D'Aloya, R., Meijer, A. J., Tager, J. M. and Quagliariello, E. (1969). α-Oxoglutarate transport in rat liver mitochondria. In *The Energy Level and Metabolic Control in Mitochondria*, pp. 159–169. Bari: Adriatica Editrice.
37. Chappell, J. B. (1969). Transport and exchange of anions in mitochondria. In *Inhibitors—Tools in Cell Research*, pp. 335–350. Heidelberg & New York: Springer-Verlag.
38. Robinson, B. H. & Chappell, J. B. (1967). An apparent energy requirement for oxaloacetate penetration into mitochondria provided by permeant cations. *Biochem. J.* **105**, 18P.
39. Palmieri, F. & Quagliariello, E. (1968). Efflux of malate of succinate from mitochondria coupled to entry of citrate. In *Mitochondria Structure and Function*. Abstr. No. 532, *Abstr. 5th FEBS Meeting*. Prague: Czechoslovak Biochemical Society.
40. Papa, S., Lofrumento, N. E., Loglisci, M. & Quagliariello, E. (1969). On the transport of inorganic phosphate and malate in rat liver mitochondria. *Biochim. biophys. Acta*, **189**, 311–314.
41. Meijer, A. J. & Tager, J. M. (1969). Effect of butylmalonate and mersalyl on anion-exchange reactions in rat liver mitochondria. *Biochim. biophys. Acta*, **189**, 136–139.
42. McGivan, J. D. & Klingenberg, M., unpublished results.
43. Azzi, A., Chappell, J. B. & Robinson, B. H. (1967). Penetration of the mitochondrial membrane by glutamate and aspartate. *Biochem. biophys. Res. Commun.* **29**, 148–152.
44. McGivan, J. D. & Chappell, J. B., unpublished work.
45. Johnson, R. N. & Chappell, J. B. (1969). The influx and efflux of phosphate in liver mitochondria. *Biochem. J.* **116**, 37P.
46. Guerin, B., Guerin, M. & Klingenberg, M. (1970). Differential inhibition of phosphate efflux and influx and a possible discrimination between an inner and outer location of the phosphate carrier in mitochondria. *FEBS Lett.*, in press.
47. Palmieri, F., personal communication.
48. Greville, G. D. & Tubbs, P. K. (1968). The catabolism of long chain fatty acids in mammalian tissues. In *Essays in Biochemistry*, vol. 4, pp. 155–212. Ed. by Campbell, P. N. & Greville, G. D. London & New York: Academic Press.
49. Beenakkers, A. M. T. & Klingenberg, M. (1964). Carnitine-coenzyme A-transacetylase in mitochondria from various organs. *Biochim. biophys. Acta*, **84**, 205–207.

50. Bremer, J., Norum, K. R. & Farstad, M. (1967). Intracellular distribution of some enzymes involved in the metabolism of fatty acids. In *Mitochondrial Structure and Compartmentation*, pp. 380–384. Bari: Adriatica Editrice.
51. Garland, P. B. & Yates, D. W. (1967). Fatty-acid oxidation. In *Mitochondrial Structure and Compartmentation*. pp. 385–399. Bari: Adriatica Editrice.
52. Tubbs, P. K. & Chase, J. F. A. (1967). Inhibition of carnitine palmitoyl-transferase by 2-bromoacyl esters of coenzyme A and carnitine. *Abstr. 4th FEBS Meeting*, p. 135. Oslo: Universitetsforlaget.
53. Beenakkers, A. M. T. & Henderson, P. T. (1967). The localization and function of carnitine acetyltransferase in the flight muscle of the locust. *Eur.J. Biochem.* **1**, 187–192.
54. Brdiczka, D., Gerbitz, K. & Pette, D. (1969). Localization and function of external and internal carnitine acetyltransferases in mitochondria of rat liver and pig kidney. *Eur. J. Biochem.* **11**, 234–240.
55. Mitchell, P. & Moyle, J. (1969). Estimation of membrane potential and pH difference across the cristae membrane of rat liver mitochondria. *Eur. J. Biochem.* **7**, 471–484.
56. Harris, E. J., Höfer, M. P. & Pressman, B. C. (1967). Stimulation of mitochondrial respiration and phosphorylation by transport inducing antibiotics. *Biochemistry, Easton*, **6**, 1348.
57. Harris, E. J. & Pressman, B. C. (1969). The direction of polarity of the mitochondrial transmembrane potential. *Biochim. biophys. Acta*, **172**, 66–70.
58. Palmieri, F., Quagliariello, E. & Klingenberg, M. (1970). Quantitative correlation between the distribution of anions and the pH difference across the mitochondrial membrane. *Eur. J. Biochem.*, in press.
59. Palmieri, F. & Quagliariello, E. (1969). Correlation between anion uptake and movement of K^+ and H^+ across the mitochondrial membrane. *Eur. J. Biochem.* **8**, 473–481.
60. Mitchell, P. (1966). *Chemiosmotic Coupling in Oxidative and Photosynthetic Phosphorylation*. Bodmin: Glynn Research Limited.
61. Mitchell, P. & Moyle, J. (1967). Respiration-driven proton translocation in rat liver mitochondria. *Biochem. J.* **105**, 1147–1162.
62. Pfaff, E., Klingenberg, M. & Heldt, H. W. (1965). Unspecific permeation and specific exchange of adenine nucleotides in mitochondria. *Biochim. biophys. Acta*, **104**, 312–315.
63. Heldt, H. W. (1969). The inhibition of adenine nucleotide translocation by atractyloside. In *Inhibitors—Tools in Cell Research*, p. 301. Heidelberg & New York: Springer-Verlag.
64. Weidemann, M. J., Erdelt, H. & Klingenberg, M. (1969). The elucidation of a carrier site for adenine nucleotide translocation in mitochondria with the help of atractyloside. In *Inhibitors—Tools in Cell Research*, pp. 324–334. Heidelberg & New York: Springer-Verlag.
65. Klingenberg, M., Grebe, K. & Pfaff, E., unpublished results.
66. Klingenberg, M. & Pfaff, E. (1968). Metabolic control in mitochondria by adenine nucleotide translocation. In *The Metabolic Roles of Citrate*. Ed. by Goodwin, T. W. *Biochem. Soc. Symp.* **27**, 105. London & New York: Academic Press.
67. Klingenberg, M., Heldt, H. W. & Pfaff, E. (1969). The role of adenine nucleotide translocation in the generation of phosphorylation energy. In *The Energy Level and Metabolic Control in Mitochondria*, p. 236. Bari: Adriatica Editrice.

68. Klingenberg, M., Wulf, R., Pfaff, E. & Heldt, H. W. (1969). Control of adenine nucleotide translocation. In *Mitochondria: Structure and Function*, pp. 59–77. Ed. by Ernster, L. & Drahota, Z. London & New York: Academic Press.
69. Kaltstein, A. M., Grebe, K. & Klingenberg, M., unpublished results.
70. Weidemann, M. J., Erdelt, H. & Klingenberg, M. (1970). Adenine nucleotide translocation in mitochondria: Identification of carrier sites, *Eur. J. Biochem.* in press.
71. Klingenberg, M., Weidemann, M. J. & Erdelt, H. (1970). Definition of carrier sites for the adenine nucleotide translocation, *Fedn Proc. Fedn Am. Socs exp. Biol.* **29**, 883.
72. Weidemann, M. J., Erdelt, H. & Klingenberg, M. (1970). Effect of bongkrekic acid on the adenine nucleotide carrier in mitochondria: Tightening of adenine nucleotide binding and differentiation between inner and outer sites. *Biochem. biophys. Res. Commun.* **39**, 363–370.

Author Index

Numbers in parentheses are reference numbers and are included to assist in locating references in the text where the authors' names are not mentioned. Numbers in italics are the pages on which the references are listed.

A

Abdullah, M., 39(77), *63*

Abeles, R. H., 39(80), 59(80), *63*

Adams, M. J., 82(53), *91*

Ambler, R. P., 28(37), *61*

Amoore, R. E., 123(4), *155*

Anderson, F. J., 30(59), *62*

Anderson, J., 126(136), *156*

Ando, K., 11(32), 12(32), *20*

Andreoli, T. E., 1(36), 12(36), *20*

Anfinsen, C. B., 71(17), 74(17), 83(58), *89*, *91*

Appleman, M. M., 26(24), 28(24), 55(24, 146), *60*, *67*

Aprile, M. A., 78(29), *90*

Arens, A., 78(35), *90*

Arion, W. J., 11(32a), *20*

Arnon, R., 105(37), *117*

Asai, J., 14(44), *21*

Ashmore, J., 84(59), 86(59), *91*

Askonas, B. A., 71(14), *89*

Assaf, S. A., 55(152), *67*

Atassi, M. Z., 105(35), 106(35), *117*

Aten, B., 71(13), 75(13), 76(26), 77(26), 78(26), 81(26), 86(26), *89*

Atkinson, D. E., 44(104), 58(104), *65*

Aune, K., 104(32), *117*

Avramovic-Zikic, O., 29(48), *62*

Azzi, A., 11(31), 16(51), *20*, *21*, 134(43), *157*

Azzone, G. F., 12(34), *20*

B

Babad, H., 95(10), *115*

Baker, E. N., 82(53), *91*

Baranowski, T., 24(6), *59*

Bartley, W., 123(4), *155*

Basu, S., 94(6), *115*

Battell, M. L. 29(47, 49), *62*

Baum, H., 9(27), 15(27), *20*

Beenakkers, A. M. T., 139(49, 53), *157*, *158*

Behrens, O. K., 85(60), *91*

Bennett, L., 88(70), *92*

Best, C. H., 82(52), *91*

Bishop, J. S., 45(105), *65*

Blake, C. C. F., 59(163), *68*, 104(30), 105(38), *117*

Blundell, T. L., 82(53), *91*

Bock, R. M., 2(3), *18*, 75(20), *89*

Bomstein, R., 9(28), *20*

Borman, A., 82(57), *91*

Boyer, P. D., 17(58), *22*

Bradbury, E. M., 104(34), *117*

Bradshaw, W. S., 81(50), *91*

Bratvold, G. E., 48(121), 49(121), 51(121, 129), *66*

Braun, V., 28(38), *61*

Brdiczka, D., 139(54), *158*

Bremer, J., 139(50), *158*

Bresler, S., 35(69), *63*

Brew, K., 96(13), 97(18), 98(13), 100(13, 18, 21, 22), 101(22), 103(22), 104(18, 31), 106(18, 31, 40), 107(41), 108(41), 109(41), 110(50), 111(21, 50, 52), 112(41, 50, 52, 53), 113(53), 114(53), *116*, *117*, *118*

Brodbeck, U., 95(11, 12), 105(36), *116*, *117*

Bromer, W. W., 75(25), 76(25), 77(25), 78(25), *89*

Brown, D. H., 24(6, 9), 41(9), *59*, *60*

Brown, N. B., 45(105), *65*

Brown, N. C., 98(15), *116*

Brown, T. H., 30(52), 35(52), *62*

Browne, W. J., 104(31), 106(31), *117*

Browning, E. T., 126(136), *156*

Bruice, T. C., 31(61), *62*

Bruni, A., 9(30), *20*

Buc, H., 41(90, 91), *64*

Buc, M. H., 41(91), *64*

de Duve, C., 81 (47, 48), 87 (47), *91*
DuVigneaud, V., 28 (41), *61*

E

Ebashi, S., 50 (126, 127), 51 (126, 127), 59 (127), *66*
Ebner, K. E., 95 (11, 12), 105 (36), *116, 117*
Elias, J. J. 108 (44), *117*
Ellis, R. M., 75 (25), 76 (25), 77 (25), 78 (25), *89*
Endo, M., 50 (127), 51 (127), 59 (127), *66*
Erdelt, H., 146 (64), 152 (64, 70), 153 (70, 71, 72), 154 (72), *158, 159*
Ernster, L., 3 (10), 12 (34), 16 (48), *19, 20, 21*
Estabrook, R. W., 3 (6), *18*
Estrada-O, S., 4 (15), *19*

F

Farstad, M., 139 (50), *158*
Fawcett, D. W., 79 (39), 80 (39), *90*
Fawcett, P., 70 (3), *88*
Fernandez-Moran, H., 3 (8), *19*
Fessenden-Raden, J. M., 7 (20), 12 (35, 38), 16 (49), *19, 20, 21*
Filmer, D., 40 (83), *63*
Findlay, J. B. C., 106 (40), *117*
Fink, C. J., 79 (42), 87 (42), *90*
Firsov, L., 35 (69), *63*
Fischer, E. H., 24 (2, 7, 10, 11), 25 (2, 18), 26 (19, 23, 24, 25, 26), 28 (24, 40), 29 (7, 25, 26, 42), 30 (7, 18, 42, 53, 54, 55, 56), 32 (7, 25, 40, 57), 33 (18, 55), 34 (18, 55, 63, 64, 65, 67), 36 (2, 25, 26, 63), 37 (26, 40, 70, 71), 38 (16, 25, 26), 39 (77), 45 (111), 46 (116, 117), 47 (117), 48 (119, 121), 49 (121), 50 (122, 123), 51 (121, 122, 123), 53 (141, 142, 143), 55 (24, 40, 70, 144, 146), 56 (144), 57 (70), *59, 60, 61, 62, 63, 65, 66, 67*
Fleischer, B., 110 (51), *118*
Fleischer, S., 110 (51), *118*
Fonyo, A., 130 (27), 131 (27), *156*
Forrey, A. W., 24 (7), 29 (7), 30 (7), 31 (7), 32 (57), *59, 62*
Fosset, M., 55 (153), 57 (153), *67*

Frerichs, H., 81 (73), 88 (71), *92*
Fried, M., 38 (74), *63*
Fujimoto, Y., 28 (39), *61*
Fukui, T., 55 (158), 57 (158), *67*

G

Garland, P. B., 139 (51), *158*
Gellert, M., 52 (135), *66*
Gerbach, E., 45 (112), *65*
Gerbitz, K., 139 (54), *158*
Gerhart, J. C., 43 (94), *64*
Givol, D., 71 (17), 74 (17), *89*
Glaser, L., 44 (98), 55 (149), 56 (98), *64, 67*
Glazer, A. N., 101 (25), *116*
Goaman, L. C. G., 104 (29), *117*
Gold, A. M., 29 (46), *62*
Goldberg, M. E., 26 (21), *60*
Goldberg, N. D., 45 (105), *65*
Goldberger, R., 9 (28), *20*
Goldberger, R. F., 71 (17), 74 (17), *89*
Gorden, P., 87 (65), 88 (65), *92*
Gottschalk, A., 94 (7), 99 (7), 100 (7), *115*
Grant, P. T., 71 (8, 12), 75 (12), 76 (12), 77 (12, 27), 78 (12), 79 (8), 81 (45, 49), 82 (49, 54), *89, 90, 91*
Gratzer, W. B., 26 (30), *61*
Graven, S. N., 4 (15), *19*
Graves, D. J., 26 (25), 29 (25), 32 (25), 36 (25), 38 (16, 25), 43 (95), 44 (97), 48 (119), 55 (152), 56 (97), *60, 64, 66, 67*
Grazer, W. B., 104 (34), *117*
Grebe, K., 130 (26), 145 (26), 147 (65), 148 (65), 151 (69), 153 (55), *156, 158, 159*
Green, A. A., 24 (1), 39 (1), *59*
Green, D. E., 2 (3, 5), *18*
Greengard, P., 52 (136), *66*
Greider, M. H., 79 (41), *90*
Greville, G. D., 16 (47), *21*, 138 (48), *157*
Grinnan, E. L., 85 (60), *91*
Grodsky, G., 88 (70), *92*
Guerin, B., 136 (46), *157*
Guerin, M., 136 (46), *157*
Guillory, R. J., 45 (108), 56 (108), 59 (108), *65*
Guirard, B. M., 29 (51), 30 (51), *62*
Gurd, F. R. N., 29 (44), *61*

Subject Index

B protein, component of lactose synthetase (*see also* α-Lactalbumin), 95–98
  activity, 100
  level in mammary gland, 107–108
  subcellular distribution of, 109–112
Butyl malonate, 130, 136

C

Calcium ions, phosphorylase kinase activation and, 49–51
Carbon dioxide transport in mitochondria, 122, 126, 127
Carbonic anhydrase (EC 4.2.1.1), 127
Carboxypeptidase B (EC 3.4.2.2), 79, 82
Carnitine, 138–139
Carriers, metabolite transport and, 120, 126, 129–133
  for adenine nucleotides, 129, 146, 151–154
  turnover number of, 146
  models of, 134–137
CF_o, 7, 12
Chemical hypothesis of oxidative phosphorylation, 13, 15–18
Chemiosmotic hypothesis of oxidative phosphorylation, 7, 11, 13, 15–18
Chymotrypsinogen, 71–72, 104
Closed space, principles of, 120–122
Collagenase (EC 3.4.4.19), 70
Conservation of conformation,
  by chymotrypsinogen and elastase, 104
  by haemoglobin and myoglobin, 104
  by α-lactalbumin and lysozyme, 103–107
Cooperative kinetics, models of, 40–41
Core protein, 9, 15
Coupling device, 12–15
Coupling factors, 7, 12–15
Crinophagy, 81
Cristae of mitochondria, 4, 5, 127
Cyclic-3, 5′-AMP, 85
  activation of phosphorylase kinase by, 49, 51–52, 54
L-Cysteine, role in PLP binding by phosphorylase, 34
Cytochrome a and a_3, location in inner mitochondrial membrane, 6–7

Cytochrome b, 7, 8, 9
Cytochrome c, 3, 6–7, 10–12
Cytochrome c_1, 7, 8, 9
Cytochrome oxidase, 1, 152
  antibody to, 11
  histochemical location of, 5, 11
  interaction of azide with, 4, 11

D

Dealanine insulin, 78
Diabetes mellitus, 70
Dicarboxylates, carrier of, 132
Di-isopropylfluorophosphate, 75
5,5′-Dithio-*bis*(2-nitrobenzoic acid) (DTNB), 136
DNA ligases, 52

E

Effectors of glycogen phosphorylase, 39–44
Elastase, 104
Endoplasmic reticulum,
  α-lactalbumin synthesis and, 111
  proinsulin synthesis and, 79, 81
Epinephrin, 45
N-Ethylmaleimide (NEM), 136
O-Ethyl-O-(*p*-nitrophenyl)-phenyl-propylphosphonate, 75

F

Ferritin, marker in study of mitochondria, 3, 5
Fetuin, terminal trisaccharide of, 99
Flash activation of phosphorylase, 53
Fructose-1,6-diphosphatase, 45

G

Galactose, 103
Galactosyl transferases, 100
Glucagon, 45, 85
Glucokinase (EC 2.7.1.1), 86
Glucosamine, 86
Glucose, control of insulin release and, 84–85

Date Due

Demco 38-297